Systems Ecology

Systems Ecology

Hillary Newland

Larsen & Keller
www.larsen-keller.com

Systems Ecology
Hillary Newland
ISBN: 978-1-64172-136-3 (Hardback)

Larsen & Keller
Published by Larsen and Keller Education,
5 Penn Plaza,
19th Floor,
New York, NY 10001, USA

Cataloging-in-Publication Data

Systems ecology / Hillary Newland.
p. cm.
Includes bibliographical references and index.
ISBN 978-1-64172-136-3
1. Ecology. 2. Ecological Systems Theory. 3. Biotic communities.
4. Bioenergetics. I. Newland, Hillary.
QH541 .S97 2019
577--dc23

Printed and bound in China.

For more information regarding Larsen and Keller Education and its products, please visit the publisher's website www.larsen-keller.com

Table of Contents

Preface

Systems ecology is a field of Earth science that studies ecological systems including ecosystems. The foundation of systems ecology lies in the approach that ecosystem is a complex system with emergent properties. It is focused on the interaction between ecological and biological systems, and the influence of human activities on the functioning of the ecosystems. Some allied fields of systems ecology are deep ecology, ecological economics, ecological energetics, ecosystem ecology, etc. This book unfolds the innovative aspects of systems ecology. The topics included herein are of the utmost significance and bound to provide incredible insights to readers. Coherent flow of topics, student-friendly language and extensive use of examples make this textbook an invaluable source of knowledge.

To facilitate a deeper understanding of the contents of this book a short introduction of every chapter is written below:

Chapter 1, The interdisciplinary field of ecology that studies the various ecological systems through a holistic approach is referred to as systems ecology. This is an introductory chapter which will discuss briefly all the significant aspects of systems ecology like ecological energetic, ecological humanities, industrial ecology, etc. **Chapter 2**, The earth system is an integrated system that comprises of the hydrosphere, biosphere, lithosphere and atmosphere. The topics elaborated in this chapter will help in developing a better perspective about these earth systems and their dynamics. **Chapter 3**, Plant Earth is constantly changing. Various cycles occurring naturally on Earth between its various spheres are called Earth cycles. This chapter discusses in detail the various types of Earth cycles, such as water cycle, carbon cycle, nitrogen cycle, rock cycle and energy cycle. **Chapter 4**, A community which consists of living organisms and non living components like air, soil and water is referred to as an ecosystem. It can be divided into various types such as aquatic ecosystem, terrestrial ecosystem, etc. The varied categories of ecosystems and their related aspects have been carefully analyzed in this chapter. **Chapter 5**, A transdisciplinary and interdisciplinary branch of academic research that addresses the coevolution and interdependence of natural ecosystems and human economies is called ecological economics. The diverse aspects of ecological economics such as green politics, natural capital, carrying capacity, degrowth, etc. have been thoroughly discussed in this chapter. **Chapter 6**, Ecological restoration can defined as the scientific study that supports the practice of renovation and restoration of degraded, destroyed or damaged ecosystems or habitats through active human intervention. This chapter discusses in detail the domains of ecological restoration such as restoration ecology, forest restoration, land restoration, etc.

Finally, I would like to thank the entire team involved in the inception of this book for their valuable time and contribution. This book would not have been possible without their efforts. I would also like to thank my friends and family for their constant support.

Hillary Newland

1

Introduction to Systems Ecology

The interdisciplinary field of ecology that studies the various ecological systems through a holistic approach is referred to as systems ecology. This is an introductory chapter which will discuss briefly all the significant aspects of systems ecology like ecological energetic, ecological humanities, industrial ecology, etc.

Systems Theory

Systems theory, also known as social systems theory, in social science is the study of society as a complex arrangement of elements, including individuals and their beliefs, as they relate to a whole (e.g., a country). The study of society as a social system has a long history in the social sciences. The conceptual origins of the approach are generally traced to the 19th century, particularly in the work of English sociologist and philosopher Herbert Spencer and French social scientist Émile Durkheim.

In the 19th century, Spencer, influenced by British naturalist Charles Darwin's theory of evolution, argued for a unitary form of the social system. In his approach, the system of society was constantly evolving into an even-more-complex state of perfection. However, alternative forms of social systems theory argued for a very different view of social evolution. In those perspectives, society is not evolving toward some perfect state; rather, it is reaching a state of increasing complexity. This was called structural differentiation. Structural differentiation refers to the adaptation of society to its environment through changes in its internal complexity.

An important aspect of social differentiation is the way in which adaptation occurs, or how changes in the structure of the system relate to the processes of the system. On one hand, society can be viewed as a total organism that is sustained by the various processes that support its function and survival. An alternative view argues that stabilizations in social systems occur not because of any rational plan of overall survival but simply because they happen to work.

Systems theory is also involved in analysing how society adapts to its environment through adjustments in its structure, with important implications for the understanding of social order. Systems theory reveals the complexity of social evolution and, on this basis, stresses the limited possibility of steering society. On the other hand, because society is vastly complex, the social scientist can nonetheless have an appreciation of the large range of adaptive possibilities for social systems.

Examples of Applications

System Dynamics

System dynamics is an approach to understanding the nonlinear behaviour of complex systems over time using stocks, flows, internal feedback loops, and time delays.

Systems Biology

Systems biology is a movement that draws on several trends in bioscience research. Proponents describe systems biology as a biology-based inter-disciplinary study field that focuses on complex interactions in biological systems, claiming that it uses a new perspective (holism instead of reduction). Particularly from the year 2000 onwards, the biosciences use the term widely and in a variety of contexts. An often stated ambition of systems biology is the modelling and discovery of emergent properties which represents properties of a system whose theoretical description requires the only possible useful techniques to fall under the remit of systems biology. It is thought that Ludwig von Bertalanffy may have created the term systems biology in 1928.

Systems Ecology

Systems ecology is an interdisciplinary field of ecology, a subset of Earth system science, that takes a holistic approach to the study of ecological systems, especially ecosystems. Systems ecology can be seen as an application of general systems theory to ecology. Central to the systems ecology approach is the idea that an ecosystem is a complex system exhibiting emergent properties. Systems ecology focuses on interactions and transactions within and between biological and ecological systems, and is especially concerned with the way the functioning of ecosystems can be influenced by human interventions. It uses and extends concepts from thermodynamics and develops other macroscopic descriptions of complex systems.

Systems Engineering

Systems engineering is an interdisciplinary approach and means for enabling the realisation and deployment of successful systems. It can be viewed as the application of engineering techniques to the engineering of systems, as well as the application of a systems approach to engineering efforts. Systems engineering integrates other disciplines and specialty groups into a team effort, forming a structured development process that proceeds from concept to production to operation and disposal. Systems engineering considers both the business and the technical needs of all customers, with the goal of providing a quality product that meets the user's needs.

Systems Psychology

Systems psychology is a branch of psychology that studies human behaviour and experience in complex systems. It received inspiration from systems theory and systems thinking, as well as the basics of theoretical work from Roger Barker, Gregory Bateson, Humberto Maturana and others. It makes an approach in psychology in which groups and individuals receive consideration as systems in homeostasis. Systems psychology "includes the domain of engineering psychology, but in addition seems more concerned with societal systems and with the study of motivational, affective, cognitive and group behavior that holds the name engineering psychology." In systems psychology, "characteristics of organizational behaviour, for example individual needs, rewards, expectations, and attributes of the people interacting with the systems, considers this process in order to create an effective system".

Developments

General Systems Research and Systems Inquiry

Many early systems theorists aimed at finding a general systems theory that could explain all systems in all fields of science. The term goes back to Bertalanffy's book, book titled "General System theory: Foundations, Development, Applications" from 1968.

Von Bertalanffy's objective was to bring together under one heading the organismic science he had observed in his work as a biologist. His desire was to use the word *system* for those principles that are common to systems in general. In GST, he writes:

> ...there exist models, principles, and laws that apply to generalized systems or their subclasses, irrespective of their particular kind, the nature of their component elements, and the relationships or "forces" between them. It seems legitimate to ask for a theory, not of systems of a more or less special kind, but of universal principles applying to systems in general.
>
> — *Von Bertalanffy*

Ervin Laszlo in the preface of von Bertalanffy's book *Perspectives on General System Theory*:

> Thus when von Bertalanffy spoke of Allgemeine Systemtheorie it was consistent with his view that he was proposing a new perspective, a new way of doing science. It was not directly consistent with an interpretation often put on "general system theory", to wit, that it is a (scientific) "theory of general systems." To criticize it as such is to shoot at straw men. Von Bertalanffy opened up something much broader and of much greater significance than a single theory (which, as we now know, can always be falsified and has usually an ephemeral existence): he created a new paradigm for the development of theories.

Ludwig von Bertalanffy outlines systems inquiry into three major domains: Philosophy, Science, and Technology. In his work with the Primer Group, Béla H. Bánáthy generalized the domains into four Integratable domains of systemic inquiry:

Domain	Description
Philosophy	the ontology, epistemology and axiology of systems
Theory	a set of interrelated concepts and principles applying to all systems
Methodology	the set of models, strategies, methods and tools that instrumentalize systems theory and philosophy
Application	the application and interaction of the domains

These operate in a recursive relationship, he explained. Integrating Philosophy and Theory as Knowledge, and Method and Application as action, Systems Inquiry then is knowledgeable action.

Cybernetics

Cybernetics is the study of the communication and control of regulatory feedback both in living and lifeless systems (organisms, organizations, machines), and in combinations of those. Its focus is how anything (digital, mechanical or biological) controls its behavior, processes information, reacts to information and changes or can be changed to better accomplish those three primary tasks.

The terms "systems theory" and "cybernetics" have been widely used as synonyms. Some authors use the term *cybernetic* systems to denote a proper subset of the class of general systems, namely those systems that include feedback loops. However Gordon Pask's differences of eternal interacting actor loops (that produce finite products) makes general systems a proper subset of cybernetics. According to Jackson (2000), von Bertalanffy promoted an embryonic form of general system theory (GST) as early as the 1920s and 1930s but it was not until the early 1950s it became more widely known in scientific circles.

Threads of cybernetics began in the late 1800s that led toward the publishing of seminal works (e.g., Wiener's *Cybernetics* in 1948 and von Bertalanffy's *General Systems Theory* in 1968). Cybernetics arose more from engineering fields and GST from biology. If anything it appears that although the two probably mutually influenced each other, cybernetics had the greater influence. Von Bertalanffy (1969) specifically makes the point of distinguishing between the areas in noting the influence of cybernetics: "Systems theory is frequently identified with cybernetics and control theory. This again is incorrect. Cybernetics as the theory of control mechanisms in technology and nature is founded on the concepts of information and feedback, but as part of a general theory of systems;" then reiterates: "the model is of wide application but should not be identified with 'systems theory' in general", and that "warning is necessary against its incautious expansion to fields for which its concepts are not made." Jackson (2000) also claims von Bertalanffy was informed by Alexander Bogdanov's three volume *Tectology* that was published in Russia between 1912 and 1917, and was translated into German in 1928. He also states it is clear to Gorelik (1975) that the "conceptual part" of general system theory (GST) had first been put in place by Bogdanov. The similar position is held by Mattessich (1978) and Capra (1996). Ludwig von Bertalanffy never even mentioned Bogdanov in his works, which Capra (1996) finds "surprising".

Cybernetics, catastrophe theory, chaos theory and complexity theory have the common goal to explain complex systems that consist of a large number of mutually interacting and interrelated parts in terms of those interactions. Cellular automata (CA), neural networks (NN), artificial intelligence (AI), and artificial life (ALife) are related fields, but they do not try to describe general (universal) complex (singular) systems. The best context to compare the different "C"-Theories about complex systems is historical, which emphasizes different tools and methodologies, from pure mathematics in the beginning to pure computer science now. Since the beginning of chaos theory when Edward Lorenz accidentally discovered a strange attractor with his computer, computers have become an indispensable source of information. One could not imagine the study of complex systems without the use of computers today.

Complex Adaptive Systems

Complex adaptive systems (CAS) are special cases of complex systems. They are *complex* in that they are diverse and composed of multiple, interconnected elements; they are *adaptive* in that they have the capacity to change and learn from experience. In contrast to control systems in which negative feedback dampens and reverses disequilibria, CAS are often subject to positive feedback, which magnifies and perpetuates changes, converting local irregularities into global features. Another mechanism, Dual-phase evolution arises when connections between elements repeatedly change, shifting the system between phases of variation and selection that reshape the system. Differently from Beer Management Cybernetics, Cultural Agency Theory (CAT) provides a modelling approach to explore predefined contexts and can be adapted to reflect those contexts.

The term *complex adaptive system* was coined at the interdisciplinary Santa Fe Institute (SFI), by John H. Holland, Murray Gell-Mann and others. An alternative conception of complex adaptive (and learning) systems, methodologically at the interface between natural and social science, has been presented by Kristo Ivanov in terms of hypersystems. This concept intends to offer a theoretical basis for understanding and implementing participation of "users", decisions makers, designers and affected actors, in the development or maintenance of self-learning systems.

Systems Ecology

Systems ecology is the analysis of how ecosystem function is determined by the components of an ecosystem and how those components cycle, retain, or exchange energy and nutrients.

Systems ecology typically involves the application of computer models that track the flow of energy and materials and predict the responses of systems to perturbations that range from fires to climate change to species extinctions. Systems ecology is closely related to mathematical ecology, with the major difference stemming from systems ecology's focus on energy and nutrient flow and its borrowing of ideas from engineering. Systems ecology is one of the few theoretical tools that can simultaneously examine a system from the level of individuals all the way up to the level of ecosystem dynamics. It is an especially valuable approach for investigating systems so large and complicated that experiments are impossible, and even observations of the entire system are impractical. In these overwhelming settings, the only approach is to break down the research into measurements of components and then assemble a system model that pieces together all components. An important contribution of ecosystem science is the recognition that there are critical ecosystem services such as cleansing of water, recycling of waste materials, production of food and fiber, and mitigation of pestilence and plagues.

Deep Ecology

Deep ecology is a somewhat recent branch of ecological philosophy (ecosophy) that considers humankind as an integral part of its environment. The philosophy emphasizes the interdependent value of human and non-human life as well as the importance of the ecosystem and natural processes. It provides a foundation for the environmental and green movements and has led to a new system of environmental ethics.

Deep ecology's core principle is the claim that, like humanity, the living environment as a whole has the same right to live and flourish. Deep ecology describes itself as "deep" because it persists in asking deeper questions concerning "why" and "how" and thus is concerned with the fundamental philosophical questions about the impacts of human life as one part of the ecosphere, rather than with a narrow view of ecology as a branch of biological science, and aims to avoid merely anthropocentric environmentalism, which is concerned with conservation of the environment only for exploitation by and for humans purposes, which excludes the fundamental philosophy of deep ecology. Deep ecology seeks a more holistic view of the world we live in and seeks to apply to life the understanding that separate parts of the ecosystem (including humans) function as a whole.

Development

The phrase "deep ecology" was coined by the Norwegian philosopher Arne Naess in 1973, and he helped give it a theoretical foundation. "For Arne Naess, ecological science, concerned with facts and logic alone, cannot answer ethical questions about how we should live. For this we need ecological wisdom. Deep ecology seeks to develop this by focusing on deep experience, deep questioning and deep commitment. These constitute an interconnected system. Each gives rise to and supports the other, whilst the entire system is, what Naess would call, an ecosophy: an evolving but consistent philosophy of being, thinking and acting in the world that embodies ecological wisdom and harmony." Naess rejected the idea that beings can be ranked according to their relative value. For example, judgments on whether an animal has an eternal soul, whether it uses reason or whether it has consciousness (or indeed higher consciousness) have all been used to justify the ranking of the human animal as superior to other animals. Næss states that from an ecological point of view "the right of all forms [of life] to live is a universal right which cannot be quantified. No single species of living being has more of this particular right to live and unfold than any other species." This metaphysical idea is elucidated in Warwick Fox's claim that we and all other beings are "aspects of a single unfolding reality". As such Deep Ecology would support the view of Aldo Leopold in his book, A Sand County Almanac that humans are "plain members of the biotic community". They also would support Leopold's "Land Ethic": "a thing is right when it tends to preserve the integrity, stability and beauty of the biotic community. It is wrong when it tends otherwise." Daniel Quinn in Ishmael, showed that an anthropocentric myth underlies our current view of the world, and a jellyfish would have an equivalent jellyfish centric view.

Deep ecology offers a philosophical basis for environmental advocacy which may, in turn, guide human activity against perceived self-destruction. Deep ecology and environmentalism hold that the science of ecology shows that ecosystems can absorb only limited change by humans or other dissonant influences. Further, both hold that the actions of modern civilization threaten global ecological well-being. Ecologists have described change and stability in ecological systems in various ways, including homeostasis, dynamic equilibrium, and "flux of nature". Regardless of which model is most accurate, environmentalists contend that massive human economic activity has pushed the biosphere far from its "natural" state through reduction of biodiversity, climate change, and other influences. As a consequence, civilization is causing mass extinction. Deep ecologists hope to influence social and political change through their philosophy.

Principles

Proponents of deep ecology believe that the world does not exist as a resource to be freely exploited by humans. If material goods do not guarantee happiness beyond a very moderate level, and over-consumption is endangering the biosphere, defining a new non-consumptive paradigm of well-being seems primordial, such a paradigm would be non-acquisitive/non-consumerist and non-hierarchical in relation to our place on Earth. The ethics of deep ecology hold that the survival of any part is dependent upon the well-being of the whole. Proponents of deep ecology offer an eight-tier platform to elucidate their claims:

- The well-being and flourishing of human and nonhuman life on Earth have value in themselves. These values are independent of the usefulness of the nonhuman world for human purposes.
- Richness and diversity of life forms contribute to the realization of these values and are also values in themselves.
- Humans have no right to reduce this richness and diversity except to satisfy vital human needs.
- The flourishing of human life and cultures is compatible with a substantial decrease of the human population. The flourishing of nonhuman life requires such a decrease.
- Present human interference with the nonhuman world is excessive, and the situation is rapidly worsening.
- Policies must therefore be changed. These policies affect basic economic, technological, and ideological structures. The resulting state of affairs will be deeply different from the present.
- The ideological change is mainly that of appreciating life quality (dwelling in situations of inherent value) rather than adhering to an increasingly higher standard of living. There will be a profound awareness of the difference between *big* and *great*.
- Those who subscribe to the foregoing points have an obligation directly or indirectly to try to implement the necessary changes.

  — Deep Ecology

These principles can be reduced to three simple propositions:

- Wilderness and biodiversity preservation
- Human population control
- Simple living (or treading lightly on the planet).

Environmental Education

Ecology in the narrow sense refers to the biological science of ecology. However, ecological paradigms and principles are being developed and applied in almost all disciplines, and these paradigms have to do with the way we approach understanding the relationships and inter-connections within and between living beings which give to each its special place and identity. Human ecology, e.g., must certainly take account of the role of our subjective lives and spiritual needs, as well as our biological ones, in terms of their ecological effects. Ecology in this sense is not a reductionist undertaking, but a movement toward a more whole (or holistic) vision and understanding of world processes. Deep ecology seeks to look into all levels of existence and might be viewed as radical by some; for them a more anthropocentric view is appropriate because it put humans at the center. Learning how to live in harmony with our surroundings is beneficial because stopping the global extinction crisis and achieving true ecological sustainability will require rethinking our values as a society. In that way education seems to be the best way to start. Sustainability education aims to help learners understand their interconnectedness with all life, to become creative problem solvers and active citizens, and to engage personally and intellectually in shaping our common future. Experiential learning and critical pedagogy are central to providing opportunities for learners to engage in transformative sustainability learning. The "Environment" broadly defined, remains somewhat neglected within development studies, despite a substantial increase in contributions to the field over the fifteen years since 2000. Several courses (with some notable exceptions) often "add on" environmental issues as special lectures or modules, and there remains a tendency for those who are grounded in the material and discursive struggles that define the discipline to consider the environment as an exotic special interest, a problem that manifests itself in societies that have the leisure to care about the natural world. Development of a modern education model promoting patriotism and civic responsibility, active social position and healthy lifestyle is closely linked to the development of environmental responsibility in the younger generation. Development of environmentally responsible personality in individual is of particular importance for graduates of educational institutions. Environmental education could be integrated in different curriculum in most fields: education for sustainable development in the context of ecopedagogy.

Ecopedagogy calls for the remaking of capitalist practices and seeks to re-engage democracy to include multispecies interests in the face of our current global ecological crisis. It does so by using different ideas that challenge the way we see education. In Critical Pedagogy, Ecoliteracy and Planetary Crisis: The Ecopedagogy Movement, Richard Kahn (2010) reformulates Herbert Marcuse's critical theories of society, and supports the kind of education that seizes the power of radical environmental activists and supports the earth democracy in which multispecies interests are represented. The destruction of habitats and threats to biodiversity resulting from expansion of human population and consumption is rarely addressed in a way that confronts students with the necessity to consider moral implications of such destruction. Pedagogically, a return to education associated with significant life experiences, such as hiking in wilderness areas as a youth; as well as strategically significant education, action competence, social learning, and variations and combinations of those and many other pedagogical approaches developed in the past 40 years. Some of these pedagogical approaches have been disputed—for example, the belief that experiencing environment first hand is an essential component of engaging people in conservation has been disputed by arguments that these education efforts have been informed by behaviorist socio-psychology models that assumed a linear causality between education experience and pro-environmental behavior. Rather, the critics have argued that people's environmental behaviors are too complex and contextually dependent to be captured by a simple casual model. The process of environmental education of schoolchildren has the following methodological characteristics:

- Goal-setting as the projected results reflects a model of environmentally responsible personality, taking into account trends in the development of key elements of education system; all natural sciences are involved in the development of basic ecological concepts.
- The introduction of interactive training methods takes place at the high school level in teaching self-reflection, hypothesizing, predicting; school natural science education is rebuilt on the basis of system approach in accordance with the planned ecologization results. Implementation of relevant methodology will promote successful development of environmentally responsible personality in high school graduates.

In higher education, the analysis of students' individual writing assignments after viewing films/documentaries presents an interesting case of using radical 'messages' within the aims of environmental education in order to trigger both student's engagement and critical thinking. The case study "If a Tree Falls and Everybody Hears the Sound" provides an example of how environmental advocacy and the objective of pluralistic education can be combined as mutually supportive means of achieving both democratic learning in which students' individual opinions are seen as extremely valuable, and simultaneously provide an example of the type of ecopedagogy that supports learning for environmental sustainability. The role of environmental advocacy can be crucially important if the interests of all planetary citizens—and not just one species—are to be taken seriously.

In her book *Wild Children — Domesticated Dreams: Civilization and the Birth of Education*, Layla AbdelRahim argues that the current institutions responsible for the construction and transmission of civilized epistemology are driven by the destructive premises at the foundation of civilization and human predatory culture. In order to return to a viable socio-environmental culture, Abdel Rahim calls for the rewilding of our anthropology (i.e. our place among other species) and of pedagogical culture, which in civilization is based on the same domestication methods of other animals.

Scientific

Old-growth forest in Biogradska Gora National Park, Montenegro.

Naess and Fox do not claim to use logic or induction to derive the philosophy directly from scientific ecology but rather hold that scientific ecology directly implies the metaphysics of deep ecology, including its ideas about the self and further, that deep ecology finds scientific underpinnings in the fields of ecology and system dynamics.

In their 1985 book *Deep Ecology*, Bill Devall and George Sessions describe a series of sources of deep ecology. They include the science of ecology itself, and cite its major contribution as the rediscovery in a modern context that "everything is connected to everything else." They point out that some ecologists and natural historians, in addition to their scientific viewpoint, have developed a deep ecological consciousness—for some a political consciousness and at times a spiritual consciousness. This is a perspective beyond the strictly human viewpoint, beyond anthropocentrism. Among the scientists they mention specifically are Rachel Carson, Aldo Leopold, John Livingston, Paul R. Ehrlich and Barry Commoner, together with Frank Fraser Darling, Charles Sutherland Elton, Eugene Odum and Paul Sears.

A further scientific source for deep ecology adduced by Devall and Sessions is the "new physics", which they describe as shattering Descartes's and Newton's vision of the universe as a machine explainable in terms of simple linear cause and effect. They propose that Nature is in a state of constant flux and reject the idea of observers as existing independent of their environment. They refer to Fritjof Capra's *The Tao of Physics* and *The Turning Point* for their characterisation of how the new physics leads to metaphysical and ecological views of interrelatedness, which, according to Capra, should make deep ecology a framework for future human societies.

The Gaia hypothesis was also an influence on the deep ecology movement.

Spiritual

The central spiritual tenet of deep ecology is that the human species is a part of the Earth, not separate from it, and as such human existence is dependent on the diverse organisms within the natural world each playing a role in the natural economy of the biosphere. Coming to an awareness of this reality involves a transformation of an outlook that presupposes humanity's superiority over the natural world. This self-realisation or "re-earthing" is used for an individual to intuitively gain an ecocentric perspective. The notion is based on the idea that the more we *expand the self* to identify with "others" (people, animals, ecosystems), the more we realize ourselves. Transpersonal psychology has been used by Warwick Fox to support this idea. Deep ecology has influenced the development of contemporary Eco spirituality.

A number of spiritual and philosophical traditions including Native American, Buddhist and Jain are drawn upon in a continuing critique of the philosophical assumptions of the modern European mind which has enabled and led to what is seen as an increasingly unsustainable level of disregard towards the rights and needs of the natural world and its ability to continue to support human life. In relation to the Judeo-Christian tradition, Naess offers the following criticism: "The arrogance of stewardship [as found in the Bible] consists in the idea of superiority which underlies the thought that we exist to watch over nature like a highly respected middleman between the Creator and Creation." This theme had been expounded in Lynn Townsend White, Jr.'s 1967 article in which however he also offered as an alternative Christian view of man›s relation to nature that of Saint Francis of Assisi, who he says spoke for the equality of all creatures, in place of the idea of man's domination over creation. Næss' further criticizes the reformation's view of creation as property to be put into maximum productive use: a view used frequently in the past to exploit and dispossess native populations. Many Protestant sects today regard the Bible's call for man to have stewardship of the earth as a call for the care for creation, rather than for exploitation.

The original Christian teachings on property support the Franciscan/stewardship interpretation of the Bible. Against this view, Martin Luther condemned church ownership of lands because "they did not want to use that property in an economically productive fashion. At best they used it to produce prayers. Luther and other Reformation leaders insisted that it should be used, not to relieve men from the necessity of working, but as a tool for making more goods. The attitude of the Reformation was practically, "not prayers, but production." And production, not for consumption, but for more production." This justification was offered to support secular takings of church endowments and properties.

Anthropologist Layla AbdelRahim sees the root of the anthropogenic degradation of the biosphere in the anthropology that constructs the human animal as the supreme predator. The ontological explanation offered for Human Supremacy by both science and religion, she says, alienate the human being from the community of life and allow for an immoral control and destruction of the wilderness, which, according to her contains the spirit and intelligence of life.

Philosophical Roots

Spinoza

Arne Naess, who first wrote about the idea of deep ecology, from the early days of developing this outlook conceived Baruch Spinoza as a philosophical source.

Others have followed Naess' inquiry, including Eccy de Jonge, in *Spinoza and Deep Ecology: Challenging Traditional Approaches to Environmentalism*, and Brenden MacDonald, in *Spinoza, Deep Ecology, and Human Diversity—Realization of Eco-Literacies*.

One of the topical centres of inquiry connecting Spinoza to Deep Ecology is "self-realization." Arne Naess in *The Shallow and the Deep, Long-Range Ecology movement* and *Spinoza and the Deep Ecology Movement* for discussion on the role of Spinoza›s conception of self-realization and its link to deep ecology.

Ecological Economics

Ecological economics (or eco-economics) is an interdisciplinary and trans disciplinary field of research that aims to integrate the earth's economic systems and its environmental health. This also includes human health and wellbeing and morals, so that they operate in a more symbiotic fashion and one does not hamper the other in terms of growth and sustainability. It aims to maintain a clear concept of the scale beyond economics and the impact we have on our environment, not necessarily for economic reasons, but reasons of conservation.

Ecological Energetics

Ecological energetics is the study of the flow of energy within an ecological system from the time the energy enters the living system until it is ultimately degraded to heat and irretrievably lost from the system.

It is also referred to as production ecology, because ecologists use the word production to describe the process of energy input and storage in ecosystems.

Ecological energetics provides information on the energetic interdependence of organisms within ecological systems and the efficiency of energy transfer within and between organisms and trophic levels. Nearly all energy enters the biota by green plants' transformation of light energy into chemical energy through photosynthesis; this is referred to as primary production. This accumulation of potential energy is used by plants, and by the animals which eat them, for growth, reproduction, and the work necessary to sustain life. The energy put into growth and reproduction is termed secondary production. As energy passes along the food chain to higher trophic levels (from plants to herbivores to carnivores), the potential energy is used to do work and in the process is degraded to heat. The laws of thermodynamics require the light energy fixed by plants to equal the energy degraded to heat, assuming the system is closed with respect to matter. An energy budget quantifies the energy pools, the directions of energy flow, and the rates of energy transformations within ecological systems.

The essentials of ecological energetics can be most readily appreciated by considering energy flowing through an individual; it is equally applicable to populations, communities, and ecosystems. Of the food energy available, only part is harvested in the process of foraging. Some is wasted, for example, by messy eaters, and the rest consumed. Part of the consumed food is transformed but is not utilized by the body, leaving as fecal material or as nitrogenous waste, the by-product of protein metabolism. The remaining energy is assimilated into the body, part of which is used to sustain the life functions and to do work—this is manifest as oxygen consumption. The remainder of the assimilated energy is used to produce new tissue, either as growth of the individual or as development of offspring. Hence production is also the potential energy (proteins, fats, and carbohydrates) on which other organisms feed. Production leads to an increase in biomass or is eliminated through death, migration, predation, or the shedding of, for example, hair, skin, and antlers.

Energy flows through the consumer food chain (from plants to herbivores to carnivores) or through the detritus food chain. The latter is fueled by the waste products of the consumer food chain, such as feces, shed skin, cadavers, and nitrogenous waste. Most detritus is consumed by microorganisms, although this food chain includes conspicuous carrion feeders like beetles and vultures. In terrestrial systems, more than 90% of all primary production may be consumed by detritus feeders. In aquatic systems, where the plants do not require tough supporting tissues, harvesting by herbivores may be efficient with little of the primary production passing to the detrivores.

Traditionally the calorie, a unit of heat energy, has been used in ecological energetics, but this has been largely replaced by the joule. Production is measured from individual growth rates and the reproductive rate of the population to determine the turnover time. The energy equivalent of food consumed, feces, and production can be determined by measuring the heat evolved on burning a sample in an oxygen bomb calorimeter, or by chemical analysis—determining the amount of carbon or of protein, carbohydrate, and lipid and applying empirically determined caloric equivalents to the values. The latter three contain, respectively, 16.3, 23.7, and 39.2 kilojoules per gram of dry weight. Maintenance costs are usually measured indirectly as respiration (normally the oxygen consumed) in the laboratory and extrapolated to the field conditions. Error is introduced by the fact that animals have different levels of activity in the field and are subject to different temperatures, and so uncertainty has surrounded these extrapolations. Oxygen consumption has been measured in animals living in the wild by using the turnover rates of doubly labeled water (D2O).

Due to the loss of usable energy with each transformation, in an area more energy can be diverted into production by plants than by consumer populations. For humans this means that utilizing plants for food directly is energetically much more efficient than converting them to eggs or meat.

Ecological Humanities

The ecological humanities (also environmental humanities) is an interdisciplinary area of research, drawing on the many environmental sub-disciplines that have emerged in the humanities over the past several decades (in particular environmental literature, environmental philosophy, environmental history and environmental anthropology). The ecological humanities aim to help bridge traditional divides between the sciences and the humanities, and between Western, Eastern and Indigenous ways of knowing the natural world and the place of humans in it.

The ecological humanities are characterised by a *connectivity ontology* and a commitment to two fundamental axioms relating to the need to submit to ecological laws and to see humanity as part of a larger living system.

Connectivity Ontology

One of the fundamental ontological presuppositions of ecological humanities is that the organic world and its inorganic parts are seen as a single system whereby each part is linked to each other part. This world view in turn shares an intimate connection with Lotka's physiological philosophy and the associated concept of the "World Engine". When we see everything as connected, then the traditional questions of the humanities concerning economic and political justice become *enlarged,* into a consideration of how justice is connected with our transformation of our environment and ecosystems. The consequence of such a connectivity ontology is, as proponents of the ecological humanities argue, that we begin to seek out a more inclusive concept of justice that includes non-humans within the domain of those to whom rights are owing. This broadened conception of justice involves "enlarged" or "ecological thinking", which presupposes the enhancement of knowledge sharing within fields of plural and diverse 'knowledges'. This kind of knowledge sharing is called trans disciplinarity. It has links with the political philosophy of Hannah Arendt and the works of Italo Calvino. As Calvino put it, "enlarge[s] the sphere of what we can imagine". It also has connections with Leibniz's Enlightenment project where, the sciences are simultaneously abridged while also being enlarged.

The situation is complicated however by the recognition of the fact that connections are both *non-linear* and *linear*. The ecological humanities therefore, require both linear and non-linear modes of language through which reasoning about justice can be done. Thus there is a motivation to find linguistic modes which can adequately express both linear and non-linear connectivities.

Axioms

There are three axioms of ecological humanities:

1. The axiom of submission to ecosystem laws;
2. The axiom of ecological kinship, which situates humanity as participant in a larger living system; and
3. The axiom of the social construction of ecosystems and ecological unity, which states that ecosystems and nature may be merely convenient conceptual entities.

Putting the first and second axioms another way, the connections between and among living things are the basis for how ecosystems are understood to work, and thus constitute laws of existence and guidelines for behaviour.

The first of these axioms has a tradition in social sciences. From the second axiom the notions of "ecological embodiment/embeddedness" and "habitat" have emerged from Political Theory with a fundamental connectivity to rights, democracy and ecologism.

The third axiom comes from the strong 'self-reflective' tradition of all 'humanities' scholarship and it encourages ecological humanities to investigate its own theoretical basis (and without which, ecological humanities is just 'ecology').

Contemporary Ideas

Political Economic Ecology

Some theorists have suggested that the inclusion of non-humans in the consideration of justice links ecocentric philosophy with political economics. This is because the theorising of justice is a central activity of political economic philosophy. If in accordance with the axioms of ecological humanities, theories of justice are enlarged to include ecological values then the necessary result is the synthesis of the concerns of ecology with that of political economy: i.e. Political Economic Ecology.

Energy Systems Language

The question of what language can best depict the linear and non-linear causal connections of ecological systems appears to have been taken up by the school of ecology known as systems ecology. To depict the linear and non-linear internal relatedness of ecosystems where the laws of thermodynamics hold significant consequences, Systems Ecologist H.T. Odum predicated the Energy Systems Language on the principles of ecological energetics. In ecological energetics, just as in ecological humanities, the causal bond between connections is considered an ontic category. Moreover, as a result of simulating ecological systems with the energy systems language H.T.O dum make the controversial suggestion that embodied energy could be understood as value, which in itself is a step into the field of Political Economic Ecology noted above.

Ecosystem Ecology

Ecosystem ecology concerns the movement of energy, water and nutrients through organisms and natural communities.

Ecosystem ecology examines large-scale ecological issues, ones that often are framed in terms not of species but rather of measures such as biomass, energy flow, and nutrient cycling. Questions include how much carbon is absorbed from the atmosphere by terrestrial plants and marine phytoplankton during photosynthesis and how much of that is consumed by herbivores, the herbivores' predators, and so on up the food chain. Carbon is the basis of life see carbon cycle, so these questions may be framed in terms of energy. How much food one has to eat each day, for instance, can be measured in terms of its dry weight or its calorie content. The same applies to measures of production for all the plants in an ecosystem or for different trophic levels of an ecosystem. A basic question in ecosystem ecology is how much production there is and what the factors is that affect it. Not surprisingly, warm, wet places such as rainforests produce more than extremely cold or dry places, but other factors are important. Nutrients are essential and may be in limited supply. The availability of phosphorus and nitrogen often determines productivity—it is the reason these substances are added to lawns and crops—and their availability is particularly important in aquatic systems. On the other hand, nutrients can represent too much of a good thing. Human activity has modified global ecosystems in ways that are increasing atmospheric carbon dioxide, a carbon source but also a greenhouse gas see greenhouse effect, and causing excessive runoff of fertilizers into rivers and then into the ocean, where it kills the species that live there.

Ecosystem Services

Ecosystem services are ecologically mediated functional processes essential to sustaining healthy human societies. Water provision and filtration, production of biomass in forestry, agriculture, and fisheries, and removal of greenhouse gases such as carbon dioxide (CO_2) from the atmosphere are examples of ecosystem services essential to public health and economic opportunity. Nutrient cycling is a process fundamental to agricultural and forest production.

However, like most ecosystem processes, nutrient cycling is not an ecosystem characteristic which can be "dialed" to the most desirable level. Maximizing production in degraded systems is an overly simplistic solution to the complex problems of hunger and economic security. For instance, intensive fertilizer use in the Midwestern United States has resulted in degraded fisheries in the Gulf of Mexico. Regrettably, a "Green Revolution" of intensive chemical fertilization has been recommended for agriculture in developed and developing countries. These strategies risk alteration of ecosystem processes that may be difficult to restore, especially when applied at broad scales without adequate assessment of impacts. Ecosystem processes may take many years to recover from significant disturbance.

For instance, large-scale forest clearance in the northeastern United States during the 18th and 19th centuries has altered soil texture, dominant vegetation, and nutrient cycling in ways that impact forest productivity in the present day. An appreciation of the importance of ecosystem function in maintenance of productivity, whether in agriculture or forestry, is needed in conjunction with plans for restoration of essential processes. Improved knowledge of ecosystem function will help to achieve long-term sustainability and stability in the poorest parts of the world.

Biomass productivity is one of the most apparent and economically important ecosystem functions. Biomass accumulation begins at the cellular level via photosynthesis. Photosynthesis requires water and consequently global patterns of annual biomass production are correlated with annual precipitation. Amounts of productivity are also dependent on the overall capacity of plants to capture sunlight which is directly correlated with plant leaf area and N content.

Net primary productivity (NPP) is the primary measure of biomass accumulation within an ecosystem. Net primary productivity can be calculated by a simple formula where the total amount of productivity is adjusted for total productivity losses through maintenance of biological processes:

$$NPP = GPP - R_{producer}$$

Where GPP is gross primary productivity and $R_{producer}$ is photosynthate (Carbon) lost via cellular respiration.

NPP is difficult to measure but a new technique known as eddy co-variance has shed light on how natural ecosystems influence the atmosphere. Figure 4 shows seasonal and annual changes in CO_2 concentration measured at Mauna Loa, Hawaii from 1987 to 1990. CO_2 concentration steadily increased, but within-year variation has been greater than the annual increase since measurements began in 1957.

These variations were thought to be due to seasonal uptake of CO_2 during summer months. A newly developed technique for assessing ecosystem NPP has confirmed seasonal variation are driven by seasonal changes in CO_2 uptake by vegetation. This has led many scientists and policy makers to speculate that ecosystems can be managed to ameliorate problems with global warming. This type of management may include reforesting or altering forest harvest schedules for many parts of the world.

Decomposition and Nutrient Cycling

Decomposition and nutrient cycling are fundamental to ecosystem biomass production. Most natural ecosystems are nitrogen (N) limited and biomass production is closely correlated with N turnover. Typically external input of nutrients is very low and efficient recycling of nutrients maintains productivity. Decomposition of plant litter accounts for the majority of nutrients recycled through ecosystems. Rates of plant litter decomposition are highly dependent on litter quality; high concentration of phenolic compounds, especially lignin, in plant litter has a retarding effect on litter decomposition. More complex C compounds are decomposed more slowly and may take many years to completely breakdown. Decomposition is typically described with exponential decay and has been related to the mineral concentrations, especially manganese, in the leaf litter.

Globally, rates of decomposition are mediated by litter quality and climate. Ecosystems dominated by plants with low-lignin concentration often have rapid rates of decomposition and nutrient cycling. Simple carbon (C) containing compounds are preferentially metabolized by decomposer microorganisms which results in rapid initial rates of decomposition, models that depend on constant rates of decay; so called "k" values, In addition to litter quality and climate, the activity of soil fauna is very important

However, these models do not reflect simultaneous linear and non-linear decay processes which likely occur during decomposition. For instance, proteins, sugars and lipids decompose exponentially, but lignin decays at a more linear rate Thus, litter decay is inaccurately predicted by simplistic models.

A simple alternative model presented in Figure 5C shows significantly more rapid decomposition that the standard model of figure 4B. Better understanding of decomposition models is an important research area of ecosystem ecology because this process is closely tied to nutrient supply and the overall capacity of ecosystems to sequester CO_2 from the atmosphere.

Trophic Dynamics

Trophic dynamics refers to process of energy and nutrient transfer between organisms. Trophic dynamics is an important part of the structure and function of ecosystems. Figure 3 shows energy transferred for an ecosystem at Silver Springs, Florida. Energy gained by primary producers (plants, P) is consumed by herbivores (H), which are consumed by carnivores (C), which are themselves consumed by "top- carnivores"(TC).

One of the most obvious patterns in Figure 3 is that as one moves up to higher trophic levels (i.e. from plants to top-carnivores) the total amount of energy decreases. Plants exert a "bottom-up" control on the energy structure of ecosystems by determining the total amount of energy that enters the system.

However, predators can also influence the structure of lower trophic levels from the top-down. These influences can dramatically shift dominant species in terrestrial and marine systems The interplay and relative strength of top-down vs. bottom-up controls on ecosystem structure and function is an important area of research in the greater field of ecology.

Trophic dynamics can strongly influence rates of decomposition and nutrient cycling in time and in space. For example, herbivory can increase litter decomposition and nutrient cycling via direct changes in litter quality and altered dominant vegetation. Insect herbivory has been shown to increase rates of decomposition and nutrient turnover due to changes in litter quality and increased frass inputs.

However, insect outbreak does not always increase nutrient cycling. Stadler showed that C rich honeydew produced during aphid outbreak can result in increased N immobilization by soil microbes thus slowing down nutrient cycling and potentially limiting biomass production. North Atlantic marine ecosystems have been greatly altered by overfishing of cod. Cod stocks crashed in the 1990s which resulted in increases in their prey such as shrimp and snow crab Human intervention in ecosystems has resulted in dramatic changes to ecosystem structure and function. These changes are occurring rapidly and have unknown consequences for economic security and human well-being.

Applications and Importance

Lessons from Two Central American Cities

The biosphere has been greatly altered by the demands of human societies. Ecosystem ecology plays an important role in understanding and adapting to the most pressing current environmental problems. Restoration ecology and ecosystem management are closely associated with ecosystem ecology. Restoring highly degraded resources depends on integration of functional mechanisms of ecosystems.

Without these functions intact, economic value of ecosystems is greatly reduced and potentially dangerous conditions may develop in the field. For example, areas within the mountainous western highlands of Guatemala are more susceptible to catastrophic landslides and crippling seasonal water shortages due to loss of forest resources. In contrast, cities such as Totonicapán that have preserved forests through strong social institutions have greater local economic stability and overall greater human well-being.

This situation is striking considering that these areas are close to each other, the majority of inhabitants are of Mayan descent, and the topography and overall resources are similar. This is a case of two groups of people managing resources in fundamentally different ways. Ecosystem ecology provides the basic science needed to avoid degradation and to restore ecosystem processes that provide for basic human needs.

Industrial Ecology

Industrial ecology conceptualises industry as a man-made ecosystem that operates in a similar way to natural ecosystems, where the waste or by product of one process is used as an input into another process. Industrial ecology interacts with natural ecosystems and attempts to move from a linear to cyclical or closed loop system. Like natural ecosystems, industrial ecology is in a continual state of flux.

Main Features

Industrial processes, from material extraction through to product disposal, have an adverse impact upon the environment. Industrial ecology aims to reduce environmental stress caused by industry whilst encouraging innovation, resource efficiency and sustained growth. Industrial ecology acknowledges that industry will continue operate and expand however, it supports industry that is environmentally conscious and has less burden upon the planet. It views industrial sites as part of a wider ecology rather than an external, solitary entity.

Within the industrial ecology concept, industry interacts with nature and utilises the wastes and by products of other industries as inputs into its own processes. Industrial ecology ranges from purely industrial ecosystems to purely natural ecosystems with a range of hybrid industrial/natural ecosystems in between. Covering both industrial management and technology, industrial ecology encompasses other sustainability concepts and tools such as material flows analysis; environmentally sound technologies; design for disassembly; and dematerialisation.

The principles of industrial ecology as defined by Tibbs (1992) are:

- Create industrial ecosystems: close the loop; view waste as a resource; create partnerships with other industries to trade by-products which are used as inputs to other processes.
- Balance industrial inputs and outputs to natural levels: manage the environmental-industrial interface; increase knowledge of ecosystem behaviour, recovery time and capacity; increase knowledge of how and when industry can interact with natural ecosystems and the limitations.
- Dematerialisation of industrial output: use less virgin materials and energy by becoming more resource efficient; reuse materials or substituting more environmentally friendly materials; do more with less.

- Improve the efficiency of industrial processes: redesign products, processes, equipment; reuse materials to conserve resources.
- Energy use - incorporate energy supply within the industrial ecology; use alternative sources of energy that have less or no impact upon the environment.
- Align policies with the industrial ecology concept: incorporate environment and economics into organisational, national and international policies; internalize the externalities; use economic instruments to encourage a move towards industrial ecology; use a more appropriate discount rate; use a more comprehensive index to measure a nation's wealth rather than GNP.

The benefits of industrial ecology include: cost savings (materials purchasing, licensing fees, waste disposal fees, etc); improved environmental protection; income generation through selling waste or by products; enhanced corporate image; improved relations with other industries and organisations and market advantages. Limitations to industrial ecology include: no market for materials; lack of support from government and industry; reluctance of industry to invest in appropriate technology; perceived legal implications and reluctance to move to another supplier.

The formation of virtual or physical eco-parks arises from clusters of industry that agree to supply or sell waste to each other, thereby moving towards the industrial ecology concept. Most eco-parks are virtual due to the high cost associated with relocating facilities. However some physical eco-parks are being designed whereby certain industries are located on the same site.

Earth Systems Science

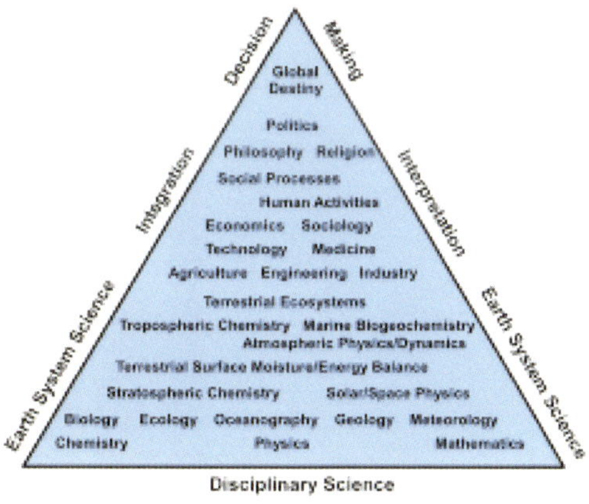

Earth system science embraces chemistry, physics, biology, mathematics and applied sciences in transcending disciplinary boundaries to treat the Earth as an integrated system and seeks a deeper understanding of the physical, chemical, biological and human interactions that determine the past, current and future states of the Earth. Earth system science provides a physical basis for understanding the world in which we live and upon which humankind seeks to achieve sustainability.

Earth system science builds upon the fundamental disciplines, which form the scientific integrating basis for any discussion of the system as a whole. This representation shows the critical relationship of ESS concepts as a bridge between the disciplines and the increasingly interpretive and integrative endeavors of human social processes. The future of our planet and the destiny of humankind are dependent upon this interdisciplinary pyramid of Earth system processes and phenomena.

Given the concerns that humankind is impacting the Earth's physical climate system, a broader concept of Earth as a system is emerging, including societal dimensions and the recognition that humanity plays an ever increasing role in global change. Within this concept, knowledge from the traditional Earth science disciplines (geology, meteorology, oceanography, ecology, biology) is being gleaned and integrated to form a physical basis for Earth System Science.

Earth system science fosters the synthesis of disciplinary knowledge into a holistic model of Earth with broader interdisciplinary relevance.

Earth Systems Engineering and Management

The biosphere, at levels from the landscape to the genome, is increasingly a product of human activity. At a landscape level, islands and mainland regions are affected by agriculture, resource extraction, human settlement, pollution, and invasive species transported by humans. Few biological communities can be found that do not reflect human predation, management, or consumption. At the organism level, species are being genetically engineered by humans to increase agricultural yields; reduce pesticide consumption; reduce demand for land for agriculture; enable plant growth under saline conditions and thereby conserve fresh water resources; produce new drugs; reduce disease; and support a healthier human diet. At the genomic level, the human genome has been mapped, as has that of selected bacteria, yeast, plants, and other mammals.

Moreover too little of the discussion about the potential effects of advancements in cutting-edge fields, such as nanotechnology, biotechnology, and information and communication technology ICT, is focused on their global impacts on integrated human-natural systems. Major human systems, from urban to economic to philosophic systems, increasingly are reflected in the physical behavior and structure of natural systems, yet there is little study and understanding of these subtle but powerful interactions.

A planet thus dominated by the activities, intentional and unintentional, of one species is a new historical phenomenon. This species is affecting a complex, dynamic system of which it is a part. Changes in such systems cannot be predicted by linear causal models; witness the continuous debate over the extent global warming is occurring, and its likely consequences. Probabilistic models and continuous data collection can help human beings enter into a dialogue with these coupled human-technological-environmental systems.

Appropriate data-gathering, modeling and dialogue is impeded by the absence of an intellectual framework within which such broad technological trends, and their cumulative impact on global human-natural systems, can be conceptualized. The current base of scientific and technical knowledge, governance institutions, and ethical approaches are inadequate to this challenge Allenby 2001. Managing these highly complex systems requires an integration of the physical and social sciences that is difficult for both cultural and disciplinary reasons, and the institutional structures that would foster this understanding, and enable its implementation, do not yet exist.

Emergence of Earth Systems Engineering and Management

The challenge of the anthropogenic Earth drove Brad Allenby to propose Earth Systems Engineering and Management ESEM, an interdisciplinary framework for perceiving, understanding, and managing complex, coupled human-natural-technological systems. It reflects not just the need to respond to, and manage, systems at scales of complexity and interconnection that current practices cannot cope with, but also to minimize the risk and scale of unplanned or undesirable perturbations in coupled human or natural systems. It does not replace traditional scientific, engineering, and social science disciplines or study; rather it draws on and integrates them to enable responsible, rational, and ethical response to the relatively new phenomenon of the anthropogenic Earth. Therefore, ESEM draws heavily on related work in multiple fields.

ESEM is a response to a broad set of multidisciplinary questions that are relatively intractable to twenty-first-century disciplinary and policy approaches: How, for example, will people cope with the potential ramifications for environmental systems of nanotechnology, biotechnology, and ICT? How can they begin to redesign human relationships with complex ecosystems such as the Everglades; engineer and manage urban centers to be more sustainable; or design Internet products and services to reduce environmental impact while increasing quality of life?

The Ethics of ESEM

Dealing responsibly with the complex web of interconnections between human and natural systems will thus require experts skilled in new approaches and frameworks, capable of creating policy and design options that protect environmental and social values while providing the desired human functionality. Such an ESEM approach requires both a rigorous understanding of the human, natural, and technological dimensions of complex systems, and an ability to design inclusive strategies to address them, all the while recognizing that no single approach or framework is likely to be able to capture the true complexity of such systems.

Even at this nascent stage, it is possible to begin to establish a set of principles applicable to ESEM:

- Try to articulate the current state of a system and desired future states, consulting with multiple stakeholders. Establish a process for continuous sharing of knowledge and revision of system goals, based on continuous monitoring of multiple system variables and their interactions. Anticipate potential problematic system responses to the extent possible, and identify markers or metrics by which shifts in probability of their occurrence may be tracked.
- The complex, information dense and unpredictable systems that are the subject of ESEM cannot be centrally or explicitly controlled. ESEM practitioners will have to be reflective, seeing themselves as an integral component of the system, closely coupled with its evolution and subject to many of its dynamics.
- Whenever possible, engineered changes should be incremental and reversible. In all cases, scale-up should allow for the fact that, especially in complex systems, discontinuities and emergent characteristics are the rule, not the exception, as scales change. Lock-in of inappropriate or untested design choices, as systems evolve over time, should be avoided.

- ESEM projects should support the evolution of system resiliency, not just redundancy. In a tightly coupled complex system, a failure of one component can be fatal, and it is virtually impossible to build in sufficient redundancy for every component Perrow 1984. The space shuttle is an example. Resilient systems are loosely coupled; the system as a whole can adapt to failures in one component. The Internet is an example, as are many natural systems. However, even in resilient systems, there are tipping points where the amount of disruption exceeds the ability of the system to adapt, and a major transformation occurs. Therefore, even resilient systems require monitoring and management.

To succeed, ESEM depends on the development of an Earth Systems Engineer ESE who would have a core area of expertise, perhaps environmental science or systems engineering or social psychology, and be able to take a global systems view of environmental problems. The ESE would have to be what Collins and Evans call an interactional expert, capable of facilitating deep, thoughtful conversations across disciplinary boundaries Collins and Evans 2002 that enable productive trading zones Galison 1997 for managing complex environmental systems Gorman and Mehalik 2002. The ESE would also be involved in the creation of new data monitoring and modeling tools that would add rigor to ESEM. To assess its value, the ESEM approach needs to be piloted on several complex systems, and the results described in detailed case-studies from which others can learn. The ESEM framework has the potential to facilitate intelligent management of trading zones centered on converging technologies: nanotechnology, biotechnology, information technology and cognition.

References

- 1968, General System theory: Foundations, Development, Applications, New York: George Braziller, revised edition 1976: ISBN 0-8076-0453-4
- Systems-theory: britannica.com, Retrieved 28 April 2018
- Anderson, Tom; Guyas, Anniina Suominen (2015). "Earth Education, Interbeing, and Deep Ecology". Studies in Art Education. 53 (3): 223–245. doi:10.1080/00393541.2012.11518865. ISSN 0039-3541.
- Deep-ecology-63: environment-ecology.com, Retrieved 25 June 2018
- Smith, Mick (2014). "Deep Ecology: What is Said and (to be) Done?". The Trumpeter. 30 (2): 141–156. ISSN 0832-6193. Retrieved 12 May 2018
- Ecological-economics-2548: safeopedia.com, Retrieved 15 May 2018
- Kopnina, Helen (2015-03-01). "If a Tree Falls and Everybody Hears the Sound: Teaching Deep Ecology to Business Students". Journal of Education for Sustainable Development. 9 (1): 101–116. doi:10.1177/0973408215569119. ISSN 0973-4082
- Earth-systems-engineering-and-management, encyclopedias-almanacs-transcripts-and-maps: encyclopedia.com, Retrieved 13 July 2018
- Blockley David, Godfrey Patrick Doing it Differently: Systems for Rethinking Infrastructure (2nd Edition) ICE Publishing, London, ISBN 978-0-7277-6082-1

2

Earth Systems

The earth system is an integrated system that comprises of the hydrosphere, biosphere, lithosphere and atmosphere. The topics elaborated in this chapter will help in developing a better perspective about these earth systems and their dynamics.

The earth system is itself an integrated system, but it can be subdivided into four main components, sub-systems or spheres: the geosphere, atmosphere, hydrosphere and biosphere. These components are also systems in their own right and they are tightly interconnected. The four main components of the earth system may be described briefly in the following way.

- The geosphere - this is the part of the planet composed of rock and minerals; it includes the solid crust, the molten mantle and the liquid and solid parts of the earth›s core. In many places, the geosphere develops a layer of soil in which nutrients become available to living organisms, and which thus provides an important ecological habitat and the basis of many forms of life. The surface of the geosphere is subject to processes of erosion, weathering and transport, as well as to tectonic forces and volcanic activity, which result in the formation of landforms such as mountains, hills and plateaux.
- The atmosphere - this is the gaseous layer surrounding the earth and held to its surface by gravity. The atmosphere receives energy from solar radiation which warms the earth›s surface and is re-emitted and conducted to the atmosphere. The atmosphere also absorbs water from the earth›s surface via the process of evaporation; it then acts to redistribute heat and moisture across the earth›s surface. In addition, the atmosphere contains substances that are essential for life, including carbon, nitrogen, oxygen and hydrogen.
- The hydrosphere - this consists of those parts of the earth system composed of water in its liquid, gaseous (vapour) and solid (ice) phases. The hydrosphere includes: the earth›s oceans and seas; its ice sheets, sea ice and glaciers; its lakes, rivers and streams; its atmospheric moisture and ice crystals; and its areas of permafrost. The hydrosphere includes both saltwater and freshwater systems, and it also includes the moisture found in the soil (soil water) and within rocks (groundwater). Water is essential for the existence and maintenance of life on earth. In some classifications, the hydrosphere is sub-divided into the fluid water systems and the cryosphere (the ice systems).
- The biosphere - this contains all living organisms and it is intimately related to the other three spheres: most living organisms require gases from the atmosphere, water from the hydrosphere and nutrients and minerals from the geosphere. Living organisms also require a medium for life, and are adapted to inhabit one or more of the other three spheres. However, much of the biosphere is contained within a shallow surface layer encompassing the lower part of the atmosphere, the surface of the geosphere and approximately the upper 100 metres of the ocean. Humans are part of the biosphere,

although they are increasingly responsible for the creation of systems that may be largely artificial (such as cities).

Hydrosphere

The detailed description of the earth systems is given below:

Hydrosphere is a discontinuous layer of water at or near Earth's surface. It includes all liquid and frozen surface waters, groundwater held in soil and rock, and atmospheric water vapour.

Water is the most abundant substance at the surface of Earth. About 1.4 billion cubic km (326 million cubic miles) of water in liquid and frozen form make up the oceans, lakes, streams, glaciers, and groundwater's found there. It is this enormous volume of water, in its various manifestations, that forms the discontinuous layer, enclosing much of the terrestrial surface, known as the hydrosphere.

Central to any discussion of the hydrosphere is the concept of the water cycle (or hydrologic cycle). This cycle consists of a group of reservoirs containing water, the processes by which water is transferred from one reservoir to another (or transformed from one state to another), and the rates of transfer associated with such processes. These transfer paths penetrate the entire hydrosphere, extending upward to about 15 km (9 miles) in Earth's atmosphere and downward to depths on the order of 5 km (3 miles) in its crust.

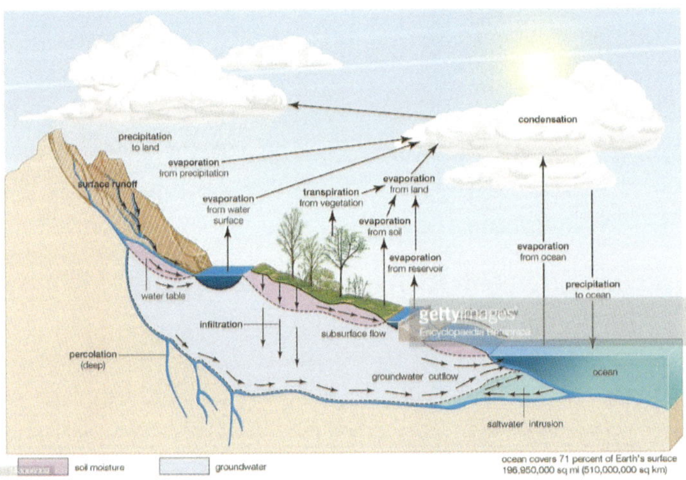

In the hydrologic cycle, water is transferred between the land surface, the ocean, and the atmosphere.

Distribution and Quantity of Earth's Waters

Ocean waters and waters trapped in the pore spaces of sediments make up most of the present-day hydrosphere. The total mass of water in the oceans equals about 50 percent of the mass of sedimentary rocks now in existence and about 5 percent of the mass of Earth's crust as a whole. Deep and shallow groundwater's constitute a small percentage of the total water locked in the pores of sedimentary rocks—on the order of 3 to 15 percent. The amount of water in the atmosphere at any one time is trivial, equivalent to roughly 13,000 cubic km (about 3,100 cubic miles) of liquid water, or about 0.001 percent of the total at Earth's surface. This water, however, plays an important role in the water cycle.

Water masses at Earth's surface		
reservoir	volume (in cubic kilometres)	percent of total
oceans	1,338,000,000	96.5
ice caps, glaciers, and permanent snow	24,064,000	1.74
ground ice and permafrost	300,000	0.22
groundwater (total)	23,400,000	1.69
groundwater (fresh)	10,530,000	0.76
groundwater (saline)	12,870,000	0.93
lakes (total)	176,400	0.013
lakes (fresh)	91,000	0.007
lakes (saline)	85,400	0.006
soil moisture	16,500	0.001
atmosphere	12,900	0.001
swamp water	11,470	0.0008
rivers	2,120	0.0002
biota	1,120	0.0001
total	1,409,560,910	101.67

At present, ice locks up a little more than 2 percent of Earth's water and may have accounted for as much as 3 percent or more during the height of the glaciations of the Pleistocene Epoch (2.6 million to 11,700 years ago). Although water storage in rivers, lakes, and the atmosphere is small, the rate of water circulation through the rain-river-ocean-atmosphere system is relatively rapid. The amount of water discharged each year into the oceans from the land is approximately equal to the total mass of water stored at any instant in rivers and lakes.

Soil moisture accounts for only 0.005 percent of the water at Earth's surface. It is this small amount of water, however, that exerts the most direct influence on evaporation from soils. The biosphere, though primarily H_2O in composition, contains very little of the total water at the terrestrial surface, only about 0.00004 percent, yet the biosphere plays a major role in the transport of water vapour back into the atmosphere by the process of transpiration.

As will be seen in the next section, Earth's waters are not pure H_2O but contain dissolved and particulate materials. Thus, the masses of water at Earth's surface are major receptacles of inorganic and organic substances, and water movement plays a dominant role in the transportation of these substances about the planet's surface.

Biogeochemical Properties of the Hydrosphere

Rainwater

About 107,000 cubic km (nearly 25,800 cubic miles) of rain fall on land each year. The total water in the atmosphere is 13,000 cubic km, and this water, owing to precipitation and evaporation, turns over every 9.6 days. Rainwater is not pure but rather contains dissolved gases and salts, fine-ground particulate material, organic substances, and even bacteria. The sources of the materials in rainwater are the oceans, soils, fertilizers, air pollution, and fossil fuel combustion.

It has been observed that rains over oceanic islands and near coasts have ratios of major dissolved constituents very close to those found in seawater. The discovery of the high salt content of rain near coastlines was somewhat surprising because sea salts are not volatile, and it might be expected that the process of evaporation of water from the sea surface would "filter" out the salts. It has been demonstrated, however, that a large percentage of the salts in rain is derived from the bursting of small bubbles at the sea surface due to the impact of rain droplets or the breaking of waves, which results in the injection of sea aerosol into the atmosphere. This sea aerosol evaporates, with resultant precipitation of the salts as tiny particles that are subsequently carried high into the atmosphere by turbulent winds. These particles may then be transported over continents to fall in rain or as dry deposition.

Assuming equilibrium with the atmospheric carbon dioxide partial pressure (P_{CO_2}) of $10^{-3.5}$ (0.00035) atmosphere, the approximate mean composition of rainwater is in parts per million (ppm): sodium (Na^+), 1.98; potassium (K^+), 0.30; magnesium (Mg^{2+}), 0.27; calcium (Ca^{2+}), 0.09; chloride (Cl^-), 3.79; sulfate (SO_4^{2-}), 0.58; and bicarbonate (HCO_3^-), 0.12. In addition to these ions, rainwater contains small amounts of dissolved silica—about 0.30 ppm. The average pH value of rainwater is 5.6. (The term pH is defined as the negative logarithm of the hydrogen ion concentration in moles per litre. The pH scale ranges from 0 to 14, with lower numbers indicating increased acidity.) On a global basis, as much as 35 percent of the sodium, 55 percent of the chlorine, 15 percent of the potassium, and 37 percent of the sulfate in river water may be derived from the oceans through sea aerosol generation.

A considerable amount of data has become available for marine aerosols. These aerosols are important because:

(1) They are vital to any description of the global biogeochemical cycle of an element,

(2) They may have an impact on climate,

(3) They are a sink, via heterogeneous chemical reactions, for trace atmospheric gases and,

they influence precipitation of cloud and rain droplets.

(4) For many trace metals, the ratio of the atmospheric flux to the riverine flux for coastal and remote oceanic areas may be greater than one, indicating the importance of atmospheric transport. Figures have been prepared that illustrate the enrichment factors (EF) of North Atlantic marine aerosols and suspended matter in North Atlantic waters relative to the crust (that is, terrestrial sources), where

$$EF_{crust} = \frac{(X|A1)_{air}}{(X|A1)_{crust}},$$

and $(X/\mathrm{Al})_{\mathrm{air}}$ and $(X/\mathrm{Al})_{\mathrm{crust}}$ refer, respectively, to the ratio of the concentration of the element X to that of Al, aluminum (which is an easily observed terrestrial component of aerosols), in the atmosphere and in average crustal material. Comparing the enrichment factors in marine aerosols with those of suspended matter in the water column indicates qualitatively the marine aerosols' importance as a source that alters the composition of marine suspended matter and, consequently, their importance to deep-sea sedimentation. Moreover, such comparisons help identify how significant terrestrial sources are for both the marine aerosols and the water below.

In some instances the ratios of ions in rainwater deviate significantly from those in seawater. Mechanisms proposed for this fractionation are, for example, the escape of chlorine as gaseous hydrogen chloride (HCl) from sea salt aerosol with a consequent enrichment in sodium and bubbling and thermal diffusion. In addition, release of biogenic gases such as dimethyl sulfide (DMS) from the sea surface and its subsequent reaction in the oceanic atmosphere to sulfate can change rainwater ion ratios with respect to seawater. Soil particles also can influence rainwater composition. Rainfall over the south western United States contains relatively high sulfate concentrations because of sulfate-bearing particles that have been blown into the atmosphere from desert soils. Rain near industrial areas commonly contains high contents of sulfate, nitrate, and carbon dioxide (CO_2) largely derived from the burning of coal and oil. There are two main processes leading to the conversion of sulfur dioxide (SO_2) to sulfuric acid (H_2SO_4). These are reactions with hydroxyl radicals (–OH) and with hydrogen peroxide (H_2O_2) in the atmosphere:

$$SO_2 + OH \rightarrow intermediate\ species \rightarrow H_2SO_4$$

and

$$SO_2 + H_2O_2 = H_2SO_4.$$

The sulfuric acid then dissociates to hydrogen and sulfate ions:

$$H_2SO_4 = 2H^+ + SO_4^{2-}.$$

For the nitrogen gases nitric oxide (NO) and nitrogen dioxide (NO_2) released from fossil fuel burning, their atmospheric reactions lead to the production of nitric acid (HNO_3) and its dissociation to hydrogen ions (H^+) and nitrate (NO_3^-). These reactions are responsible for the acid rain conditions that occurred in the north-eastern United States, south-eastern Canada, and western Europe during the second half of the 20th century. The high sulfate values of the rain in the north-eastern United States reflect the acid precipitation conditions of this region.

Eutrophication

Discharge of Denver sewage treatment plant in the Platte River (USA).

Eutrophication is the (mostly undesired) increase in the concentration of nutrients to an ecosystem. Increased nutrient enrichment can arise from both point and non-point sources:

- Point source pollution: Pollution that comes from contaminants that enter a waterway from a single identifiable source such as stationary locations or fixed facilities. Examples are discharges from a sewage treatment plant or industrial plants and fish farms.
- Non-point source pollution: Pollution from widespread including human activities with no specific point of discharge or entry into receiving water courses. Examples are leaching out of nitrogen compounds from fertilized agricultural lands and losses from atmospheric deposition.

Manure run-off, an example of non-point source pollution.

The enrichment of water by nutrients can be of a natural origin (natural eutrophication) but is often dramatically increased by human activities (cultural or anthropogenic eutrophication). Natural eutrophication has been occurring for millennia. It is the process of addition, flow and accumulation of nutrients to water bodies resulting in changes to the primary production and species composition of the community. Cultural eutrophication is the process that speeds up natural eutrophication because of human activity. There are three main sources of anthropogenic nutrient input: erosion and leaching from fertilized agricultural areas, and sewage from cities and industrial waste water. Atmospheric deposition of nitrogen (from animal breeding and combustion gases) can also be important.

The most common nutrients causing eutrophication are nitrogen and phosphorus. The main source of nitrogen pollutants is run-off from agricultural land, whereas most phosphorus pollution comes from households and industry, including phosphorus-based detergents. These nutrients enter aquatic ecosystems via the air, surface water or groundwater. Most of the commercially fixed nitrogen and mined phosphorus goes into production of fertilizer. The rising demand for fertilizer has come from the need to meet the nutritional demands of our rapidly expanding human population. The rise in intensive fertilizer use has serious implications for coastal habitats because greater application results in greater runoff, and the fraction of fertilizer lost from fields will increase with intensity of application. Increased global productions of nitrogenous fertilizers have largely been linked to concerns over the relationship between water quality and eutrophication. Nutrient removal in sewage treatment plants and promotion of phosphorus-free detergents are vital to minimize the impact of nitrogen and phosphorus pollution in Europe's waters.

Ecological Effects

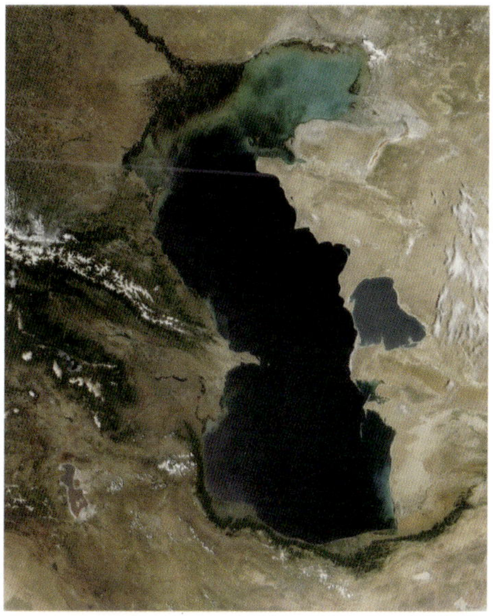

Eutrophication is apparent as increased turbidity in the northern part of the Caspian Sea, imaged from orbit.

Eutrophication was recognized as a water pollution problem in European and North American lakes and reservoirs in the mid-20th century. Since then, it has become more widespread. Surveys showed that 54% of lakes in Asia are eutrophic; in Europe, 53%; in North America, 48%; in South America, 41%; and in Africa, 28%. In South Africa, a study by the CSIR using remote sensing has shown more than 60% of the dams surveyed were eutrophic. Some South African scientists believe that this figure might be higher with the main source being dysfunctional sewage works that produce more than 4 billion liters a day of untreated, or at best partially treated, sewage effluent that discharges into rivers and dams.

Many ecological effects can arise from stimulating primary production, but there are three particularly troubling ecological impacts: decreased biodiversity, changes in species composition and dominance, and toxicity effects.

- Increased biomass of phytoplankton
- Toxic or inedible phytoplankton species
- Increases in blooms of gelatinous zooplankton

- Increased biomass of benthic and epiphytic algae
- Changes in macrophyte species composition and biomass
- Decreases in water transparency (increased turbidity)
- Colour, smell, and water treatment problems
- Dissolved oxygen depletion
- Increased incidences of fish kills
- Loss of desirable fish species
- Reductions in harvestable fish and shellfish
- Decreases in perceived aesthetic value of the water body

Decreased Biodiversity

When an ecosystem experiences an increase in nutrients, primary producers reap the benefits first. In aquatic ecosystems, species such as algae experience a population increase (called an algal bloom). Algal blooms limit the sunlight available to bottom-dwelling organisms and cause wide swings in the amount of dissolved oxygen in the water. Oxygen is required by all aerobically respiring plants and animals and it is replenished in daylight by photosynthesizing plants and algae. Under eutrophic conditions, dissolved oxygen greatly increases during the day, but is greatly reduced after dark by the respiring algae and by microorganisms that feed on the increasing mass of dead algae. When dissolved oxygen levels decline to hypoxic levels, fish and other marine animals suffocate. As a result, creatures such as fish, shrimp, and especially immobile bottom dwellers die off. In extreme cases, anaerobic conditions ensue, promoting growth of bacteria such as *Clostridium botulinum* that produces toxins deadly to birds and mammals. Zones where this occurs are known as dead zones.

New Species Invasion

Eutrophication may cause competitive release by making abundant a normally limiting nutrient. This process causes shifts in the species composition of ecosystems. For instance, an increase in nitrogen might allow new, competitive species to invade and out-compete original inhabitant species. This has been shown to occur in New England salt marshes. In Europe and Asia, the common carp frequently lives in naturally Eutrophic or Hyper eutrophic areas, and is adapted to living in such conditions. The eutrophication of areas outside its natural range partially explain the fish's success in colonising these areas after being introduced.

Toxicity

Some algal blooms, otherwise called "nuisance algae" or "harmful algal blooms", are toxic to plants and animals. Toxic compounds they produce can make their way up the food chain, resulting in animal mortality. Freshwater algal blooms can pose a threat to livestock. When the algae die or are eaten, neuro- and hepatotoxins are released which can kill animals and may pose a threat to humans. An example of algal toxins working their way into humans is the case of shellfish poisoning.

Biotoxins created during algal blooms are taken up by shellfish (mussels, oysters), leading to these human foods acquiring the toxicity and poisoning humans. Examples include paralytic, neurotoxic, and diarrhoetic shellfish poisoning. Other marine animals can be vectors for such toxins, as in the case of ciguatera, where it is typically a predator fish that accumulates the toxin and then poisons humans.

Prevention and Reversal

Eutrophication poses a problem not only to ecosystems, but to humans as well. Reducing eutrophication should be a key concern when considering future policy, and a sustainable solution for everyone, including farmers and ranchers, seems feasible. While eutrophication does pose problems, humans should be aware that natural runoff (which causes algal blooms in the wild) is common in ecosystems and should thus not reverse nutrient concentrations beyond normal levels. Cleanup measures have been mostly, but not completely, successful. Finnish phosphorus removal measures started in the mid-1970s and have targeted rivers and lakes polluted by industrial and municipal discharges. These efforts have had a 90% removal efficiency. Still, some targeted point sources did not show a decrease in runoff despite reduction efforts.

Shellfish in Estuaries: Unique Solutions

One proposed solution to eutrophication in estuaries is to restore shellfish populations, such as oysters and mussels. Oyster reefs remove nitrogen from the water column and filter out suspended solids, subsequently reducing the likelihood or extent of harmful algal blooms or anoxic conditions. Filter feeding activity is considered beneficial to water quality by controlling phytoplankton density and sequestering nutrients, which can be removed from the system through shellfish harvest, buried in the sediments, or lost through denitrification. Foundational work toward the idea of improving marine water quality through shellfish cultivation was conducted by Odd Lindahl, using mussels in Sweden. In the United States, shellfish restoration projects have been conducted on the East, West and Gulf coasts. Nutrient pollution for an extended explanation of nutrient remediation using shellfish.

Minimizing Nonpoint Pollution Future Work

Nonpoint pollution is the most difficult source of nutrients to manage. The literature suggests, though, that when these sources are controlled, eutrophication decreases. The following steps are recommended to minimize the amount of pollution that can enter aquatic ecosystems from ambiguous sources.

Riparian Buffer Zones

Studies show that intercepting non-point pollution between the source and the water is a successful means of prevention. Riparian buffer zones are interfaces between a flowing body of water and land, and have been created near waterways in an attempt to filter pollutants; sediment and nutrients are deposited here instead of in water. Creating buffer zones near farms and roads is another possible way to prevent nutrients from traveling too far. Still, studies have shown that the effects of atmospheric nitrogen pollution can reach far past the buffer zone. This suggests that the most effective means of prevention is from the primary source.

Prevention Policy

Laws regulating the discharge and treatment of sewage have led to dramatic nutrient reductions to surrounding ecosystems, but it is generally agreed that a policy regulating agricultural use of fertilizer and animal waste must be imposed. In Japan the amount of nitrogen produced by livestock is adequate to serve the fertilizer needs for the agriculture industry. Thus, it is not unreasonable to command livestock owners to clean up animal waste—which when left stagnant will leach into ground water.

Policy concerning the prevention and reduction of eutrophication can be broken down into four sectors: Technologies, public participation, economic instruments, and cooperation. The term technology is used loosely, referring to a more widespread use of existing methods rather than an appropriation of new technologies. As mentioned before, nonpoint sources of pollution are the primary contributors to eutrophication, and their effects can be easily minimized through common agricultural practices. Reducing the amount of pollutants that reach a watershed can be achieved through the protection of its forest cover, reducing the amount of erosion leeching into a watershed. Also, through the efficient, controlled use of land using sustainable agricultural practices to minimize land degradation, the amount of soil runoff and nitrogen-based fertilizers reaching a watershed can be reduced. Waste disposal technology constitutes another factor in eutrophication prevention. Because a major contributor to the nonpoint source nutrient loading of water bodies is untreated domestic sewage, it is necessary to provide treatment facilities to highly urbanized areas, particularly those in underdeveloped nations, in which treatment of domestic waste water is a scarcity. The technology to safely and efficiently reuse waste water, both from domestic and industrial sources, should be a primary concern for policy regarding eutrophication.

The role of the public is a major factor for the effective prevention of eutrophication. In order for a policy to have any effect, the public must be aware of their contribution to the problem, and ways in which they can reduce their effects. Programs instituted to promote participation in the recycling and elimination of wastes, as well as education on the issue of rational water use are necessary to protect water quality within urbanized areas and adjacent water bodies.

Economic instruments, "which include, among others, property rights, water markets, fiscal and financial instruments, charge systems and liability systems, are gradually becoming a substantive component of the management tool set used for pollution control and water allocation decisions." Incentives for those who practice clean, renewable, water management technologies are an effective means of encouraging pollution prevention. By internalizing the costs associated with the negative effects on the environment, governments are able to encourage a cleaner water management.

Because a body of water can have an effect on a range of people reaching far beyond that of the watershed, cooperation between different organizations is necessary to prevent the intrusion of contaminants that can lead to eutrophication. Agencies ranging from state governments to those of water resource management and non-governmental organizations, going as low as the local population, are responsible for preventing eutrophication of water bodies. In the United States, the most well known inter-state effort to prevent eutrophication is the Chesapeake Bay.

Nitrogen Testing and Modelling

Soil Nitrogen Testing (N-Testing) is a technique that helps farmers optimize the amount of fertilizer applied to crops. By testing fields with this method, farmers saw a decrease in fertilizer application costs, a decrease in nitrogen lost to surrounding sources, or both. By testing the soil and modeling the bare minimum amount of fertilizer are needed, farmers reap economic benefits while reducing pollution.

Organic Farming

There has been a study that found that organically fertilized fields "significantly reduce harmful nitrate leaching" over conventionally fertilized fields. However, a more recent study found that eutrophication impacts are in some cases higher from organic production than they are from conventional production.

Dredging

Dredging helps in reduction of soil contamination that was caused by sewage sludge (like fecal sludge), wilted plants, or chemical spill.

Bokashi

Bokashi is a modern method of eutrophication prevention to decrease smell and toxic materials. Bokashi is Japanese for "shading off" or "gradation." It derives from the practice of Japanese farmers centuries ago of covering food waste with rich, local soil that contained the microorganisms that would ferment the waste. After a few weeks, they would bury the waste. Bokashi balls used in water pollution are made up of soil (either clay or loam or mixture of both), water, molasses, and antimicrobial solution; antimicrobial solution is used to kill germs (either bacteria or fungi), that cause foul odor, and algae.

Acid Rain

Acid rain refers to a mixture of deposited material, both wet and dry, coming from the atmosphere containing more than normal amounts of nitric and sulfuric acids. Simply put, it means rain that is acidic in nature due to the presence of certain pollutants in the air due to cars and industrial processes. It is easily defined as rain, fog, sleet or snow that has been made acidic by pollutants in the air as a result of fossil fuel and industrial combustions that mostly emits Nitrogen Oxides (NOx) and Sulfur Dioxide (SO_2). Acidity is determined on the basis of the pH level of the water droplets. Normal rain water is slightly acidic with a pH range of 5.3-6.0, because carbon dioxide and water present in the air react together to form carbonic acid, which is a weak acid. When the pH level of rain water falls below this range, it becomes acid rain.

When these gases react with water molecules and oxygen among other chemicals found in the atmosphere, mild acidic chemical compounds such as sulfuric and nitric acid are formed resulting to acid rain. Acid rain generally leads to weathering of buildings, corrosion of metals, and peeling of paints on surfaces. Erupting volcanoes contains some chemicals that can cause acid rain. Apart from this, burning of fossil fuels, running of factories and automobiles due to human activities are few other reasons behind this activity.

Presently, large amounts of acid deposition is witnessed in the southeastern Canada, northeastern United States and most of Europe, including portions of Sweden, Norway, and Germany. In addition, some amount of acid deposition is found in parts of South Asia, South Africa, Sri Lanka, and Southern India.

Forms of Acid Rain

There are two forms in which acid deposition occurs – wet and dry. Both are discussed below:

Wet Deposition: When the wind blows the acidic chemicals in the air to the areas where the weather is wet, the acids fall to the ground in the form of rain, sleet, fog, snow or mist. It removes acid from the atmosphere and deposit them on the earth's surface. When this acid flows through the ground, it affects large number of plants, animals and aquatic life. The water from drain flows into rivers and canals which is them mixed up with sea water, thereby affecting marine habitats.

Dry Deposition: If the wind blows the acidic chemicals in the air to the areas where the weather is dry, the acidic pollutants slip into dust or smoke and fall to the ground as dry particles. These stick to the ground and other surfaces such as cars, houses, trees and buildings. Almost 50% of the acidic pollutants in the atmosphere fall back through dry deposition. These acidic pollutants can be washed away from earth surface by rainstorms.

It was discovered way back in 1800s during the Industrial Revolution. A Scottish chemist, Robert Angus Smith, was first to discover this phenomenon in 1852 as a relationship between acid rain and atmospheric pollution in Manchester, England. But it gained public attention mainly in 1960s. The term was coined in 1972 when the NY Times published reports about the climate change effects which started arising due to the occurrence of acid rain in the Hubbard Brook Experimental Forest in New Hampshire.

Causes of Acid Rain

Both natural and man-made sources are known to play a role in the formation of acid rain. But, it is mainly caused by combustion of fossil fuels which results in emissions of sulfur dioxide (SO2) and nitrogen oxides (NOx).

Natural Sources

The major natural causal agent for acid rain is volcanic emissions. Volcanoes emit acid producing gases to create higher than normal amounts of acid rain or any other form of precipitation such as fog and snow to an extent of affecting vegetation cover and health of residents within the surrounding. Decaying vegetation, wildfires and biological processes within the environment also generate the acid rain forming gases. Dimethly sulfide is a typical example of a major biological contributor to sulfur containing elements into the atmosphere. Lighting strikes also naturally produces nitric oxides that react with water molecules via electrical activity to produce nitric acid, thereby forming acid rain.

Man-made Sources

Human activities leading to chemical gas emissions such as sulfur and nitrogen are the primary contributors to acid rain. The activities include air pollution sources emitting sulfur and nitrogen gases like factories, power generations facilities, and automobiles. In particular, use of coal for electrical power generation is the biggest contributor to gaseous emissions leading to acid rain. Automobiles and factories also release high scores of gaseous emissions on daily basis into the air, especially in highly industrialized areas and urban regions with large numbers of car traffic. These gases react in the atmosphere with water, oxygen, and other chemicals to form various acidic compounds such as sulfuric acid, ammonium nitrate, and nitric acid. As a result, these areas experience exceedingly high amounts of acid rain.

The existing winds blow these acidic compounds over large areas across borders and they fall back to the ground in the form of acid rain or other forms of precipitation. Upon reaching the earth, it flows across the surface, absorbs into the soil and enters into lakes and rivers and finally gets mixed up with sea water.

The gases i.e. i.e. sulfur dioxide (SO2) and nitrogen oxides (NOx) are primarily gases occurring from electric power generation by burning coal and responsible for acid rain.

Effects of Acid Rain

Acid rain has significant effects on the world environment and public health.

- Effect on Aquatic Environment: Acid rain either falls directly on aquatic bodies or gets run off the forests, roads and fields to flow into streams, rivers and lakes. Over a period of time, acids get accumulated in the water and lower the overall pH of the water body. The aquatic plants and animals need a particular pH level of about 4.8 to survive. If the pH level falls below that the conditions become hostile for the survival of aquatic life. Acid rain tendency of altering pH and aluminum concentrations greatly affects pH concentration levels in surface water, thereby affecting fish as well as other aquatic life-forms. At pH levels below 5, most fish eggs cannot hatch. Lower pHs can also kill adult fish. Acid rain runoff from catchment areas into rivers and lakes has also reduced biodiversity as rivers and lakes become more acidic. Species including fish, plant and insect types in some lakes, rivers and brooks have been reduced and some even completely eliminated owing to excess acid rain flowing into the waters.

- Effect on Forests: It makes trees vulnerable to disease, extreme weather, and insects by destroying their leaves, damaging the bark and arresting their growth. Forest damage due to acid rain is most evident in Eastern Europe – especially Germany, Poland and Switzerland.
- Effect on Soil: Acid rain highly impacts on soil chemistry and biology. It means, soil microbes and biological activity as well as soil chemical compositions such as soil pH are damaged or reversed due to the effects of acid rain. The soil needs to maintain an optimum pH level for the continuity of biological activity. When acid rains seep into the soil, it means higher soil pH, which damages or reverses soil biological and chemical activities. Hence, sensitive soil microorganisms that cannot adapt to changes in pH are killed. High soil acidity also denatures enzymes for the soil microbes. On the same breadth, hydrogen ions of acid rain leach away vital minerals and nutrients such as calcium and magnesium.
- Vegetation Cover and Plantations: The damaging effects of acid rain on soil and high levels of dry depositions have endlessly damaged high altitude forests and vegetation cover since they are mostly encircled by acidic fogs and clouds. Besides, the widespread effects of acid rain on ecological harmony have lead to stunted growth and even death of some forests and vegetation cover.
- Effect on Architecture and Buildings: Acid rain on buildings, especially those constructed with limestone, react with the minerals and corrode them away. This leaves the building weak and susceptible to decay. Modern buildings, cars, airplanes, steel bridges and pipes are all affected by acid rain. Irreplaceable damage can be caused to the old heritage buildings.
- Effect on Public Health: When in atmosphere, sulfur dioxide and nitrogen oxide gases and their particulate matter derivatives like sulfates and nitrates, degrades visibility and can cause accidents, leading to injuries and deaths. Human health is not directly affected by acid rain because acid rain water is too dilute to cause serious health problems. However, the dry depositions also known as gaseous particulates in the air which in this case are nitrogen oxides and sulfur dioxide can cause serious health problems when inhaled. Intensified levels of acid depositions in dry form in the air can cause lung and heart problems such as bronchitis and asthma.

- Other Effects: Acid rain leads to weathering of buildings, corrosion of metals, and peeling of paints on surfaces. Buildings and structures made of marble and limestone are the ones especially damaged by acid rain due to the reactivity of the acids in the rain and the calcium compounds in the structures. The effects are commonly seen on statues, old grave stones, historic monuments, and damaged buildings. Acid rain also corrodes metals like steel, bronze, copper, and iron.

Solutions to Acid Rain

Cleaning up Exhaust Pipes and Smokestacks

Most of the electric power supporting the modern-day energy requirements comes from combusting fossil fuels such as oil, natural gas, and coal that generate nitrogen oxides (NOx) and sulfur dioxide (SO2) as the chief contributors to acid rain. Burning coal largely accounts for SO2 emissions while NOx emissions are mostly from fossil fuel combustions.

Washing coal, use of coal comprised of low sulfur, and use of devices known as “scrubbers” can provide technical solution to SO2 emissions. “Scrubbing” also called flue-gas desulfurization (FGD) typically work to chemically eliminate SO2 from the gases leaving smokestacks. It can eliminate up to 95% of SO2 gases. Power generation facilities can also shift to using fuels that emit much less SO2 such as natural gas instead of burning coal. These methods are simply called emission reduction strategies.

Similarly, NOx emissions from automobile fossil fuel combustions are mitigated upon by use of catalytic converters. Catalytic converters are fixed on the exhaust pipe system to reduce NOx emission. Improvement of gasoline that combusts cleaner is also a strategy for reducing emission of NOx gases.

Restoring Damaged Environments

Use of limestone or lime, a process called liming, is a practice that people can do to repair the damage caused by acid rain to lakes, rivers and brooks. Adding lime into acidic surface waters balances the acidity. It’s a process that has extensively been used, for instance in Sweden, to keep the water pH at optimum. Even though, liming is an expensive method and has to be done repeatedly. Furthermore, it only offers a short-term solution at the expense of solving the broader challenges of SO2 and NOx emissions and risks to human health. Nevertheless, it helps to restore and allow the survival of aquatic life forms by improving chronically acidified surface waters.

Alternative Energy Sources

Besides fossil fuels, there is a wide range of alternative energy sources that can generate electrical power. These include wind energy, geothermal energy, solar energy, hydropower, and nuclear power. Harnessing these energy sources can offer effective electrical power alternatives instead of using fossil fuels. Fuel cells, natural gas, and batteries can also substitute use of fossil fuel as cleaner energy sources. As of today, all energy sources have environmental and economic costs as well as benefits. The only solution is using sustainable energy that can protect the future.

Individual, National/State, and International Actions

Millions of people directly and indirectly contribute to SO2 and NOx emissions. Mitigation of this challenge requires individuals to be more informed about energy conservation and ways of reducing emissions such as: turning off lights or electrical appliances when not using them; use public transport; use energy efficient electrical appliances; and use of hybrid vehicles or those with low NOx emissions.

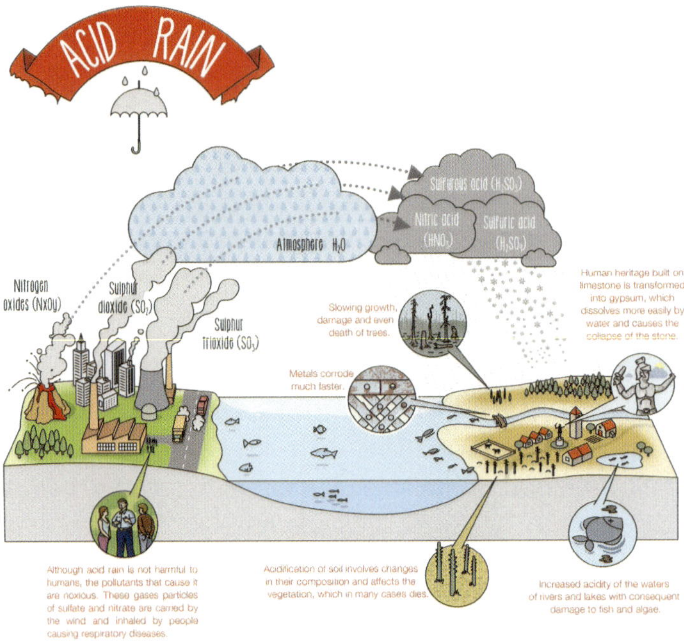

Biosphere

Biosphere is a relatively thin life-supporting stratum of Earth's surface, extending from a few kilometres into the atmosphere to the deep-sea vents of the ocean. The biosphere is a global ecosystem composed of living organisms (biota) and the abiotic (nonliving) factors from which they derive energy and nutrients.

Before the coming of life, Earth was a bleak place, a rocky globe with shallow seas and a thin band of gases—largely carbon dioxide, carbon monoxide, molecular nitrogen, hydrogen sulfide, and water vapour. It was a hostile and barren planet. This strictly inorganic state of the Earth is called the geosphere; it consists of the lithosphere (the rock and soil), the hydrosphere (the water), and the atmosphere (the air). Energy from the Sun relentlessly bombarded the surface of the primitive Earth, and in time—millions of years—chemical and physical actions produced the first evidence of life: formless, jellylike blobs that could collect energy from the environment and produce more of their own kind. This generation of life in the thin outer layer of the geosphere established what is called the biosphere, the "zone of life," an energy-diverting skin that uses the matter of the Earth to make living substance.

The biosphere is a system characterized by the continuous cycling of matter and an accompanying flow of solar energy in which certain large molecules and cells are self-reproducing. Water is a major predisposing factor, for all life depends on it. The elements carbon, hydrogen, nitrogen, oxygen, phosphorus, and sulfur, when combined as proteins, lipids, carbohydrates, and nucleic acids, provide the building blocks, the fuel, and the direction for the creation of life. Energy flow is required to maintain the structure of organisms by the formation and splitting of phosphate bonds.

Organisms are cellular in nature and always contain some sort of enclosing membrane structure, and all have nucleic acids that store and transmit genetic information.

All life on Earth depends ultimately upon green plants, as well as upon water. Plants utilize sunlight in a process called photosynthesis to produce the food upon which animals feed and to provide, as a by-product, oxygen, which most animals require for respiration. At first, the oceans and the lands were teeming with large numbers of a few kinds of simple single-celled organisms, but slowly plants and animals of increasing complexity evolved. Interrelationships developed so that certain plants grew in association with certain other plants, and animals associated with the plants and with one another to form communities of organisms, including those of forests, grasslands, deserts, dunes, bogs, rivers, and lakes. Living communities and their nonliving environment are inseparably interrelated and constantly interact upon each other. For convenience, any segment of the landscape that includes the biotic and abiotic components is called an ecosystem. A lake is an ecosystem when it is considered in totality as not just water but also nutrients, climate, and all of the life contained within it. A given forest, meadow, or river is likewise an ecosystem. One ecosystem grades into another along zones termed ecotones, where a mixture of plant and animal species from the two ecosystems occurs. A forest considered as an ecosystem is not simply a stand of trees but is a complex of soil, air, and water, of climate and minerals, of bacteria, viruses, fungi, grasses, herbs, and trees, of insects, reptiles, amphibians, birds, and mammals.

Stated another way, the abiotic, or nonliving, portion of each ecosystem in the biosphere includes the flow of energy, nutrients, water, and gases and the concentrations of organic and inorganic substances in the environment. The biotic, or living, portion includes three general categories of organisms based on their methods of acquiring energy: the primary producers, largely green plants; the consumers, which include all the animals; and the decomposers, which include the microorganisms that break down the remains of plants and animals into simpler components for recycling in the biosphere. Aquatic ecosystems are those involving marine environments and freshwater environments on the land. Terrestrial ecosystems are those based on major vegetation types, such as forest, grassland, desert, and tundra. Particular kinds of animals are associated with each such plant province.

Ecosystems may be further subdivided into smaller biotic units called communities. Examples of communities include the organisms in a stand of pine trees, on a coral reef, and in a cave, a valley, a lake, or a stream. The major consideration in the community is the living component, the organisms; the abiotic factors of the environment are excluded.

A community is a collection of species populations. In a stand of pines, there may be many species of insects, of birds, of mammals, each a separate breeding unit but each dependent on the others for its continued existence. A species, furthermore, is composed of individuals, single functioning units identifiable as organisms. Beyond this level, the units of the biosphere are those of the organism: organ systems composed of organs, organs of tissues, tissues of cells, cells of molecules, and molecules of atomic elements and energy. The progression, therefore, proceeding upward from atoms and energy, is toward fewer units, larger and more complex in pattern, at each successive level.

The Diversity of Life

The biosphere supports between 3 and 30 million species of plants, animals, fungi, single-celled prokaryotes such as bacteria, and single-celled eukaryotes such as protozoans (Figure . Of this total, only about 1.4 million species have been named so far, and fewer than 1 percent have been studied for their ecological relationships and their role in ecosystems. A little more than half the named species are insects, which dominate terrestrial and freshwater communities worldwide; the laboratories of systematists are filled with insect species yet to be named and described. Hence, the relationships of organisms to their environments and the roles that species play in the biosphere are only beginning to be understood.

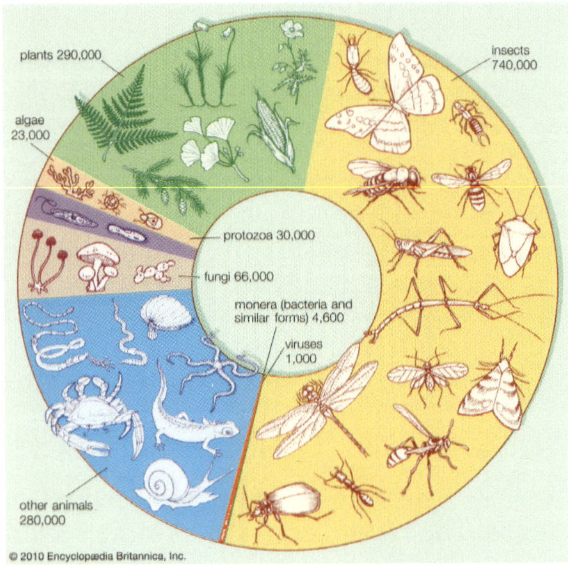

Figure: Estimated number of known living species. The majority of species are still unknown—*i.e.*, yet to be described by taxonomists.

The Organization of the Biosphere

Natural Groupings

This tremendous diversity of life is organized into natural ecological groupings. As life has evolved, populations of organisms have become separated into different species that are reproductively isolated from one another. These species are organized through their interrelationships into complex biological communities. The interactions in these communities affect, and are affected by, the physical environments in which they occur, thereby forming ecosystems through which the energy and nutrients necessary for life flow and cycle. The mix of species and physical environments vary across the globe, creating ecological communities, or biomes, such as the boreal forests of North America and Eurasia and the rainforests of the tropics. The sum total of the richness of these biomes is the biosphere.

Processes of Evolution

This hierarchical organization of life has come about through the major processes of evolution—natural selection (the differential success of the reproduction of hereditary variations resulting from the interaction of organisms with their environment), gene flow (the movement of genes among different populations of a species), and random genetic drift (the genetic change that occurs in small populations owing to chance). Natural selection operates on the expressed characteristics of genetic variants found within populations, winnowing members of the population who are less well suited to their environment from those better suited to it. In this manner, populations become adapted to their local ecosystems, which include both the physical environment and the other species with which they interact in order to survive and reproduce.

The genetic variation that is necessary for a species to adapt to the physical environment and to other organisms arises from new mutations within populations, the recombination of genes during sexual reproduction, and the migration of and interbreeding with individuals from other populations. In very small populations, however, some of that variation is lost by chance alone through random genetic drift. The combined result of these evolutionary processes is that after many generations populations of the same species have widely divergent characteristics. Some of these populations eventually become so genetically different that their members cannot successfully interbreed, resulting in the evolution of a separate species (speciation).

The diversification of life through local adaptation of populations and speciation has created the tremendous biodiversity found on the Earth. In most regions a square kilometre (0.4 square mile) will harbour hundreds—in some places even thousands—of species. The interactions between these species create intricate webs of relationships as the organisms reciprocally evolve, adapting to one another and becoming specialized for their interactions. Natural communities of species reflect the sum of these species' interactions and the ongoing complex selection pressures they constantly endure that drive their evolution. The many ecological and evolutionary processes that affect the relationships among species and their environments render ecology one of the most intricate of the sciences. The answers to the major questions in ecology require an understanding of the relative effects of many variables acting simultaneously.

Closed Ecological Systems

Biosphere 2

Closed ecological systems (CES) are ecosystems that do not rely on matter exchange with any part outside the system.

The term is most often used to describe small manmade ecosystems. Such systems are scientifically interesting and can potentially serve as a life support system during space flights, in space stations or space habitats.

In a closed ecological system, any waste products produced by one species must be used by at least one other species. If the purpose is to maintain a life form, such as a mouse or a human, waste products such as carbon dioxide, feces and urine must eventually be converted into oxygen, food, and water.

A closed ecological system must contain at least one autotrophic organism. While both chemotrophic and phototrophic organisms are plausible, almost all closed ecological systems to date are based on a phototroph such as green algae.

Examples

A closed ecological system for an entire planet is called an ecosphere.

Man-made closed ecological systems which were created to sustain human life include Biosphere 2, MELiSSA, and the BIOS-1, BIOS-2, and BIOS-3 projects.

Bottle gardens and aquarium ecospheres are partially or fully enclosed glass containers that are self-sustaining closed ecosystems that can be made or purchased. They can include tiny shrimp, algae, gravel, decorative shells, and gorgonia.

Lithosphere

The lithosphere is the solid and rigid outer layer of our planet. It includes the crust and part of the upper mantle that contains rigid rocks. Beneath this layer is the asthenosphere where the rocks in this part of the upper mantle are not rigid. The rocks can flow like a liquid or break apart similar to silly putty.

Two Types of Crustal Plates

This layer of the Earth contains two very different types of crust. The continental crust contains a variety of rocks. They are igneous rocks, sedimentary rocks and metamorphic rocks that make up the rock cycle. Continental crust is lighter than oceanic crust which is made of basalt and gabbro. These rocks are derived from the upper mantle.

Oceanic Crust

The oceanic crust is much younger because it is constantly be created at spreading zones and recycled in subduction zones. New oceanic crust forms when crustal plates separate. Molten rock from the upper mantle that has collected in magma chambers oozes onto the ocean floor forming a layer of rocks between the spreading plates. This is the newest and youngest crust on the surface of the Earth.

Continental Crust

The lithosphere contains the continental crust that is much older than oceanic plates. Continental plates in subduction zones are not recycled because they override the oceanic plates. If continental plates meet that create a collision zone and great mountain chains like the Himalayas rise above the landscape.

Continental Plate Collides with Oceanic Plate

When a continental plate and an oceanic plate meet the continental plate overrides the oceanic plate. As the continental plate overrides the oceanic plate it scrapes off the top layers off the oceanic plate.

San Francisco Terranes

The layers that are scraped off the oceanic plates are called terranes. Seventeen of these accreted terranes have been identified in the San Francisco bay area and are related to the movement along the San Andreas Fault.

Crustal Plate Boundaries

Scientists use earthquakes to determine the boundaries of the crustal plates on the surface of the Earth. Earthquakes along fault lines occur frequently as plates move around the Earth.

Major Crustal Plates

The seven major plates that contain the bulk of the continents and the Pacific Ocean: African Plate, Antarctic Plate, Eurasian Plate, Indo-Australian Plate, North American Plate, South American Plate and the Pacific Plate.

Smaller Plates

There are also eight smaller plates that are generally shown on maps that display crust plates. These are the smaller but very important plates that have cause great earthquakes in the past.

- Arabian Plate
- Caribbean Plate
- Cocos Plate, Indian Plate

- Juan de Fuca Plate
- Nazca Plate
- Philippine Sea Plate
- Scotia Plate.

Scientists have also identified a number of micro-plates where the land is being torn apart or are subducting.

Interaction of Lithosphere with other Spheres

The cool, brittle lithosphere is just one of five great "spheres" that shape the environment of Earth. The other spheres are the biosphere (Earth's living things); the cryosphere (Earth's frozen regions, including both ice and frozen soil); the hydrosphere (Earth's liquid water); and the atmosphere (the air surrounding our planet). These spheres interact to influence such diverse elements as ocean salinity, biodiversity, and landscape.

For instance, the pedosphere is part of the lithosphere made of soil and dirt. The pedosphere is created by the interaction of the lithosphere, atmosphere, cryosphere, hydrosphere, and biosphere. Enormous, hard rocks of the lithosphere may be ground down to powder by the powerful movement of a glacier (cyrosphere). Weathering and erosion caused by wind (atmosphere) or rain (hydrosphere) may also wear down rocks in the lithosphere. The organic components of the biosphere, including plant and animal remains, mix with these eroded rocks to create fertile soil—the pedosphere.

The lithosphere also interacts with the atmosphere, hydrosphere, and cryosphere to influence temperature differences on Earth. Tall mountains, for example, often have dramatically lower temperatures than valleys or hills. The mountain range of the lithosphere is interacting with the lower air pressure of the atmosphere and the snowy precipitation of the hydrosphere to create a cool or even icy climate zone. A region's climate zone, in turn, influences adaptations necessary for organisms of the region's biosphere.

The rocky lithosphere includes part of the upper mantle and crust.

The LAB

The depth of the lithosphere-asthenosphere boundary (LAB) is a hot topic among geologists and rheologists. These scientists study the upper mantle's viscosity, temperature, and grain size of its rocks and minerals. What they have found varies widely, from a thin, crust-deep boundary at mid-ocean ridges to thick, 200-meter (124-mile) boundary beneath cratons, the oldest and most stable parts of continental lithosphere.

Atmosphere

Atmosphere is the gas and aerosol envelope that extends from the ocean, land, and ice-covered surface of a planet outward into space. The density of the atmosphere decreases outward, because the gravitational attraction of the planet, which pulls the gases and aerosols (microscopic suspended particles of dust, soot, smoke, or chemicals) inward, is greatest close to the surface. Atmospheres of some planetary bodies, such as Mercury, are almost non-existent, as the primordial atmosphere has escaped the relatively low gravitational attraction of the planet and has been released into space. Other planets, such as Venus, Earth, Mars, and the giant outer planets of the solar system, have retained an atmosphere. In addition, Earth's atmosphere has been able to contain water in each of its three phases (solid, liquid, and gas), which has been essential for the development of life on the planet.

The atmosphere is comprised of layers based on temperature. These layers are the troposphere, stratosphere, mesosphere and thermosphere. A further region at about 500 km above the Earth's surface is called the exosphere.

The Different Layers of the Atmosphere

The atmosphere can be divided into layers based on its temperature, as shown in the figure below. These layers are the troposphere, the stratosphere, the mesosphere and the thermosphere. A further region, beginning about 500 km above the Earth's surface, is called the exosphere.

The red line on the figure below shows how temperature varies with height. The scale on the right shows the pressure. For example, at a height of 50 km, the pressure is only about one thousandth of the pressure at the ground.

The Troposphere

This is the lowest part of the atmosphere - the part we live in. It contains most of our weather - clouds, rain, and snow. In this part of the atmosphere the temperature gets colder as the distance above the earth increases, by about 6.5 °C per kilometre. The actual change of temperature with height varies from day to day, depending on the weather.

The troposphere contains about 75% of all of the air in the atmosphere, and almost all of the water vapour (which forms clouds and rain). The decrease in temperature with height is a result of the decreasing pressure. If a parcel of air moves upwards it expands (because of the lower pressure). When air expands it cools. So air higher up is cooler than air lower down.

The lowest part of the troposphere is called the boundary layer. This is where the air motion is determined by the properties of the Earth's surface. Turbulence is generated as the wind blows over the Earth's surface, and by thermals rising from the land as it is heated by the sun. This turbulence redistributes heat and moisture within the boundary layer, as well as pollutants and other constituents of the atmosphere.

The top of the troposphere is called the tropopause. This is lowest at the poles, where it is about 7 - 10 km above the Earth's surface. It is highest (about 17 - 18 km) near the equator.

The Stratosphere

This extends upwards from the tropopause to about 50 km. It contains much of the ozone in the atmosphere. The increase in temperature with height occurs because of absorption of ultraviolet (UV) radiation from the sun by this ozone. Temperatures in the stratosphere are highest over the summer pole, and lowest over the winter pole.

By absorbing dangerous UV radiation, the ozone in the stratosphere protects us from skin cancer and other health damage. However chemicals (called CFCs or freons, and halons) which were once used in refrigerators, spray cans and fire extinguishers have reduced the amount of ozone in the stratosphere, particularly at polar latitudes, leading to the so-called "Antarctic ozone hole".

Now humans have stopped making most of the harmful CFCs we expect the ozone hole will eventually recover over the 21st century, but this is a slow process.

The Mesosphere

The region above the stratosphere is called the mesosphere. Here the temperature again decreases with height, reaching a minimum of about -90°C at the "mesopause".

The Thermosphere and Ionosphere

The thermosphere lies above the mesopause, and is a region in which temperatures again increase with height. This temperature increase is caused by the absorption of energetic ultraviolet and X-Ray radiation from the sun.

The region of the atmosphere above about 80 km is also caused the "ionosphere", since the energetic solar radiation knocks electrons off molecules and atoms, turning them into "ions" with a positive charge. The temperature of the thermosphere varies between night and day and between the seasons, as do the numbers of ions and electrons which are present. The ionosphere reflects and absorbs radio waves, allowing us to receive shortwave radio broadcasts in New Zealand from other parts of the world.

The Exosphere

The region above about 500 km is called the exosphere. It contains mainly oxygen and hydrogen atoms, but there are so few of them that they rarely collide - they follow "ballistic" trajectories under the influence of gravity, and some of them escape right out into space.

The Magnetosphere

The earth behaves like a huge magnet. It traps electrons (negative charge) and protons (positive), concentrating them in two bands about 3,000 and 16,000 km above the globe - the Van Allen "radiation" belts. This outer region surrounding the earth, where charged particles spiral along the magnetic field lines, is called the magnetosphere.

References

- Schindler, David and Vallentyne, John R. (2004) Over fertilization of the World's Freshwaters and Estuaries, University of Alberta Press, p. 1, ISBN 0-88864-484-1
- Hydrosphere, science: britannica.com, Retrieved 22 Mayl 2018
- Whiteside, M. C. (1983). "The mythical concept of eutrophication". Hydrobiologia. 103: 107–150. doi:10.1007/BF00028437
- Causes-and-effects-of-acid-rain: conserve-energy-future.com, Retrieved 12 April 2018
- "Critical Loads – Atmospheric Deposition". U.S Forest Service. United States Department of Agriculture. Retrieved 2 April 2018
- What-is-acid-rain, climate-change: activesustainability.com, Retrieved 30 June 2018
- Angold P. G. (1997). "The Impact of a Road Upon Adjacent Heathland Vegetation: Effects on Plant Species Composition". The Journal of Applied Ecology. 34 (2): 409–417. doi:10.2307/2404886. JSTOR 2404886
- Biosphere, science: britannica.com, Retrieved 31 March 2018
- Callisto, Marcos; Molozzi, Joseline and Barbosa, José Lucena Etham (2014) "Eutrophication of Lakes" in A. A. Ansari, S. S. Gill (eds.), Eutrophication: Causes, Consequences and Control, Springer Science+Business Media Dordrecht. doi:10.1007/978-94-007-7814-6_5. ISBN 978-94-007-7814-6
- lithosphere, encyclopedia: nationalgeographic.org, Retrieved 19 May 2018
- Smith, V. H.; Tilman, G. D.; Nekola, J. C. (1999). "Eutrophication: Impacts of excess nutrient inputs on freshwater, marine, and terrestrial ecosystems". Environmental Pollution (Barking, Essex : 1987). 100 (1–3): 179–196. doi:10.1016/S0269-7491(99)00091-3. PMID 15093117

3

Understanding Earth Cycles

Plant Earth is constantly changing. Various cycles occurring naturally on Earth between its various spheres are called Earth cycles. This chapter discusses in detail the various types of Earth cycles, such as water cycle, carbon cycle, nitrogen cycle, rock cycle and energy cycle.

The earth cycles include the water cycle, carbon cycle, nitrogen cycle, rock cycle and the energy cycle.

Water Cycle

Water cycle, also called hydrologic cycle, cycle that involves the continuous circulation of water in the Earth-atmosphere system. Of the many processes involved in the water cycle, the most important are evaporation, transpiration, condensation, precipitation, and runoff. Although the total amount of water within the cycle remains essentially constant, its distribution among the various processes is continually changing.

A brief treatment of the water cycle follows. For full treatment, hydrosphere: The water cycle.

Evaporation, one of the major processes in the cycle, is the transfer of water from the surface of the Earth to the atmosphere. By evaporation, water in the liquid state is transferred to the gaseous, or vapour, state. This transfer occurs when some molecules in a water mass have attained sufficient kinetic energy to eject themselves from the water surface. The main factors affecting evaporation are temperature, humidity, wind speed, and solar radiation. The direct measurement of evaporation, though desirable, is difficult and possible only at point locations. The principal source of water vapour is the oceans, but evaporation also occurs in soils, snow, and ice. Evaporation from snow and ice, the direct conversion from solid to vapour, is known as sublimation. Transpiration is the evaporation of water through minute pores, or stomata, in the leaves of plants. For practical purposes, transpiration and the evaporation from all water, soils, snow, ice, vegetation, and other surfaces are lumped together and called evapotranspiration, or total evaporation.

Water vapour is the primary form of atmospheric moisture. Although its storage in the atmosphere is comparatively small, water vapour is extremely important in forming the moisture supply for dew, frost, fog, clouds, and precipitation. Practically all water vapour in the atmosphere is confined to the troposphere (the region below 6 to 8 miles [10 to 13 km] altitude).

The transition process from the vapour state to the liquid state is called condensation. Condensation may take place as soon as the air contains more water vapour than it can receive from a free water surface through evaporation at the prevailing temperature. This condition occurs as the consequence of either cooling or the mixing of air masses of different temperatures. By condensation, water vapour in the atmosphere is released to form precipitation.

Precipitation that falls to the Earth is distributed in four main ways: some is returned to the atmosphere by evaporation, some may be intercepted by vegetation and then evaporated from the surface of leaves, some percolates into the soil by infiltration, and the remainder flows directly as surface runoff into the sea. Some of the infiltrated precipitation may later percolate into streams as groundwater runoff. Direct measurement of runoff is made by stream gauges and plotted against time on hydrographs.

Most groundwater is derived from precipitation that has percolated through the soil. Groundwater flow rates, compared with those of surface water, are very slow and variable, ranging from a few millimetres to a few metres a day. Groundwater movement is studied by tracer techniques and remote sensing.

Ice also plays a role in the water cycle. Ice and snow on the Earth's surface occur in various forms such as frost, sea ice, and glacier ice. When soil moisture freezes, ice also occurs beneath the Earth's surface, forming permafrost in tundra climates. About 18,000 years ago glaciers and ice caps covered approximately one-third of the Earth's land surface. Today about 12 percent of the land surface remains covered by ice masses.

Stages of Water Cycle

Different Stages of the Water Cycle

Water goes through three different states in the water cycle. It can be a liquid (water), a gas (water vapor) or a solid (ice). These three states are interchangeable, as water can freeze into ice or evaporate into water vapor, water vapor can condense as water, and ice can melt into water. The water cycle consists of a number of steps which sees water go through each of these states.

Evaporation: Water is found in lakes, oceans, swamps, and soil, as well as in all living creatures and plants. When heat is applied from the sun, through exertion, or by artificial means, the water molecules become excited and spread out. The loss of density is called 'evaporation', and it sees the water rise into the air forming clouds of water vapor. Normally, the evaporation of water occurs when the water hits boiling point, around one hundred degrees centigrade.

However, in places in which the air pressure and humidity is lower, far less heat energy is needed to evaporate the water because there is less pressure holding the water molecules together. The water that evaporates from the oceans is not salty, as the salt is too dense and heavy to rise with the water vapor, which is why water from rivers and lakes is not salty.

- Snow and ice can actually turn into water vapor without first turning into water. This process is called 'sublimation', and it results from low humidity and dry winds. This usually occurs at the peaks of mountains or other high-up places, as the lower air pressure means that less energy is needed to sublimate the ice into water vapor.

Some of the highest peaks on earth, such as Mount Everest, have all of the necessary components for sublimation, namely: strong sunlight, low temperatures, low air pressure, strong wind, and low humidity. If you've ever seen dry ice, which has a fog pouting off of it, this is an example of sublimation in action.

- When water evaporates off of a plant's leaves, the process is known as 'evapotranspiration'. A large percentage of the water in the atmosphere is produced by this process due to the large areas covered by plants and trees across the planet. While about ninety per cent of the water vapor in the world comes from lakes, oceans, and streams, the remaining ten per cent is comprised of the various plant life around the world.

Condensation: The water vapor that has risen into the sky cools significantly when it comes into contact with the cooler air found up high. The vapor becomes a cloud, which is pushed around the world by moving air currents and winds.

- If the water vapor cools to anything above zero degrees centigrade, it will condense as water. Essentially, the water vapor will start to condense on the surface of tiny particles of dust and dirt that rose with the vapor during the process of evaporation. These tiny droplets will start to fall into one another and merge, producing a larger droplet. When a droplet is large enough, gravity will pull it down at a rate that exceeds the updraft in the cloud, leading to the droplet falling out of the cloud and onto the ground below. This process is called 'precipitation', or – more commonly – rainfall.
- If precipitation occurs in conditions which are particularly cold or have very low air pressure, then these water droplets can quite often crystallize and freeze. This causes the water to fall as solid ice, known as hail, or as snow. If the conditions are in between those associated with snow and rain, the droplets will fall as icy cold, half frozen water commonly referred to as sleet.

Precipitation: The water that has fallen as rain is absorbed into the ground through a process known as 'infiltration'. Soil and other porous materials can absorb great deals of water this way, while rocks and other harder substances will only retain a small amount of water.

When the water infiltrates soil, it will move in all directions until it either seeps into nearby streams or else sink deeper into what is known as 'groundwater storage'. This is where the water that does not seep out or evaporate joins up under the ground, saturating the smallest nooks and crannies of rock and soil under the ground. These formations are also known as 'aquifers', and explain why sometimes the ground underneath the top soil is damp or sodden.

- When an aquifer becomes too full, it starts to leak out onto the surface forming what is known commonly as a 'spring'. These can often be found in formations of porous or brittle rock, which can crack following slightly acidic rainfall. Should the water be located near a volcano or any source of natural thermal energy, it will form a hot spring.

Runoff: After the water has fallen and the soil has become saturated, or the snow has melted, the water follows gravity and falls down any hills, mountains, or other inclines to form or join rivers. This process is known as 'runoff', and it is how water comes to rest in lakes and returns to the ocean. The water falls according to the incline of the place from which it is falling, and when several threads of water meet they form a stream.

The direction in which the water moves is known as 'stream flow', and it is central to the concept of the currents within rivers and streams. These streams and rivers will run off eventually to either form lakes or re-join the ocean, depending on their proximity to the ocean. Due to the amount of water stored in snow or ice, sudden increase in the heat can lead to flooding due to the water suddenly melting and running off at an alarming rate. This is why flooding can occur so easily during a warm spring following a particularly cold and biting winter.

When more snow falls than evaporates or sublimates, the ice will compact densely to form what are known as 'ice caps'. The ice caps and glaciers located in the coldest regions of the world are the biggest collections of ice in the world, and are slowly starting to shrink as the water in which they sit is becoming warmer.

This happens in a cyclical manner, with no beginning or end. As precipitation happens in one part of the world evapotranspiration is happening somewhere else. The cycle of water never ends, which is why the oceans stay level, there are always clouds somewhere in the sky, and drinkable water doesn't suddenly just run out. At some point, any water that has infiltrated will be released, any that has joined the ocean will evaporate, and even a large portion of the ice caps may melt, releasing water that has been stored for millions of years back into the water cycle with potentially cataclysmic effects.

Human Impact on Water Cycle

Urbanisation

One of the most concerning human activities that affects the entire water cycle is urbanisation. This happens when the natural water cycle cannot function properly in urban areas due to buildings, concrete and other surfaces that are preventing the water from reaching the ground, allowing it to soak into the soil.

The ability to keep trees, plants and grass healthy also starts to decrease, as low soil moisture hinders the healthy growth of plants. This means that irrigation is needed, which, in turn, causes water wastage.

When humans remove natural vegetation and replace those areas with infrastructure, it speeds up overflow which leads to evaporation and higher river levels. The overflow water that we're left with is storm water, which cannot be consumed. Storm water contains pollutants such as sediment, nutrients, bacteria and chemicals that can threaten human and animal health.

Land Cover Changes

Land-cover changes are changes that are directly influenced by local, regional or global climate processes, whereas land-use changes are changes that are affected by humans.

Land-use changes can be anything from evolving economic and social to biophysical conditions, and land-cover changes are seen as the global wind patterns and topography, which both play a major role in the moisture recycling and distribution patterns. Although we cannot control these elements, we can control land-use factors (like irrigations, artificial dams and deforestation) which all increase evaporation patterns, affecting runoff water, the yearly stream flow and the vegetation from the land.

Artificial Reservoirs

As much as artificial reservoirs are great sources for water conservation, they have a negative effect on the water cycle. These artificial water sources, often referred to as man-made reservoirs, can be formed by building a dam across a valley, diverting river flow into the reservoir. Although they help to save water, they cause more environmental stress on land and have a big impact on time concentration of watersheds because of the evaporation caused. Because these water storage units are used for drinking water, storm water lands in these reservoirs which causes the water to be contaminated. If you do drink this water, make sure that you purify it first to protect your health.

Climate Change

Much like many other environmental challenges, climate change is one of the most obvious effects on the entire water cycle. We already know that the cycle is driven by sunlight radiation and the global temperature distribution, which in turn affects the distribution of rainfall. The more humans influence the quality of the available water, the bigger the effect is on global patterns of stream flow, meaning that changes in sea level, ocean salinity and in biophysical properties will result in rapid climate change.

Infrastructure and building will cause conventional methods of land development to collect and convey storm water quickly into a series of drains and pipes that will flow directly into the closest water body or man-made artificial reservoir, with little or no water quality treatment.

Deforestation

This is the process that is referred to when clearing all or removing specific trees from forest land. ‘Deforestation’ happens when builders turn this land into non-forest use.

When we remove trees from forests that have been growing for years, it reduces evapotranspiration, which is the sum of evaporation and plant transpiration from land and ocean surface to the atmosphere, leading to a possible reduction in precipitation. This then leads to an increase in water eutrophication, which then cannot be consumed by humans.

Cloud Seeding

Cloud seeding is a weather modification, where you change the amount or type of precipitation that falls from clouds through the usage of harmful substances. During this process, the substances that fall from the clouds are dispersed into the air, causing cloud condensation which further affects climate conditions.

Carbon Cycle

The carbon cycle is the cycle by which the element carbon moves through our Earth's various systems. It is a fascinating and complex process because living things, atmospheric changes, ocean chemistry, and geologic activity are all part of this cycle.

The Earth is a closed system; that is, it cannot gain or lose materials or resources. Instead, materials must be used and re-used many times over for life to be sustainable. This applies to materials including carbon, oxygen, nitrogen, and water.

Carbon is an essential element for life as we know it because of its ability to form multiple, stable bonds with other molecules.

This is why nucleotides, amino acids, sugars, and lipids all depend on carbon backbones: carbon provides a stable structure that allows the chemistry of life to happen. Without carbon, none of these molecules could exist and function in the ways that permit the chemistry of life to occur.

Our type of life is referred to as "carbon-based life" by scientists. Some scientists speculate that life on other planets could be based on a different chemistry, but no evidence such life has yet been found.

Originally, carbon was created through nuclear fusion in the hearts of stars. When those stars exploded, materials including carbon were scattered through space. And when the Earth was formed – it was formed of some of those carbon-containing materials.

In time, living things developed as a result of carbon and other atoms reacting in the chemically volatile atmosphere of ancient Earth. When some of these molecules became self-replicating, they began to take carbon from the atmosphere and even from rocks and use it to build materials for life such as sugars, proteins, and lipids.

Early photosynthesizers such as cyanobacteria transformed Earth's atmosphere by turning huge amounts of carbon dioxide gas into oxygen gas. This was not their goal – but rather a happy side effect of these first life forms removing the "C" from CO_2. This left O_2 in the atmosphere – and countless molecules of carbon serving as part of the machinery for life.

In time, other forms of life discovered that the new O_2 in the atmosphere could be used to power a highly efficient method of liberating the energy stored in carbon-based organic molecules like sugars. This allowed these life forms to use the energy plants used to make these molecule for their own metabolism.

Using this process of "cellular respiration," animals and other oxygen-breathers started turning O_2 back into CO_2 – effectively spitting out the carbon atoms once contained in sugars, proteins, and lipids after extracting all of their energy.

This happy balance of plants turning carbon dioxide into living matter while animals reduce it back into gases for the plants to consume has existed for billions of years. In the process, new steps became incorporated – such as the formation of fossil fuels, which occurs when organic matter such as dead plants and animals become trapped underground by geologic processes.

The graphic below illustrates some common ways in which carbon moves through the ecosystem:

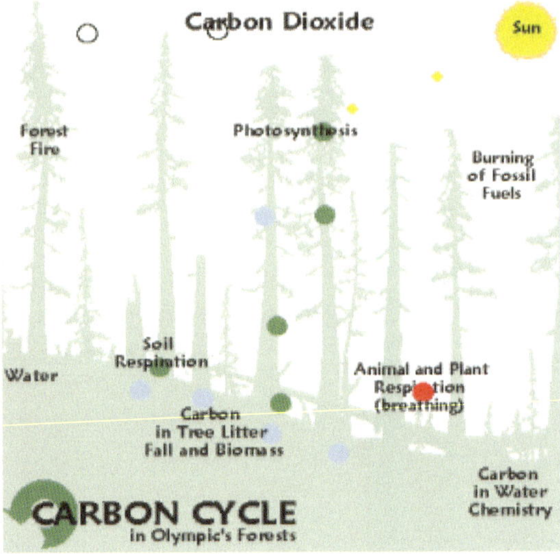

Recently, humans have made some big changes to the Earth's carbon cycle.

By burning huge amounts of fossil fuels and cutting down roughly half of the Earth's forests, humans have decreased the Earth's ability to take carbon out of the atmosphere, while releasing large amounts of carbon into the atmosphere that had been stored in solid form as plant matter and fossil fuels.

This means more carbon dioxide in Earth's atmosphere – which is particularly dangerous since carbon dioxide is a "greenhouse gas" that plays a role in regulating the Earth's temperature and weather patterns.

Function of Carbon Cycle

The carbon cycle, under normal circumstances, works to ensure the stability of variables such as the Earth's atmosphere, the acidity of the ocean, and the availability of carbon for use by living things.

Each of its components is of crucial importance to the health of all living things – especially humans, who rely on many food crops and animals to feed our large population.

Carbon dioxide in the atmosphere prevents the sun's eat from escaping into space, very much like the glass walls of a greenhouse. This isn't always a bad thing – some carbon dioxide in the atmosphere is good for keeping the Earth warm and its temperature stable.

But Earth has experienced catastrophic warming cycles in the past, such as the Permian extinction, which is thought to have been caused by a drastic increase in the atmosphere's level of greenhouse gases.

No one is sure what caused the change that brought about the Permian extinction. Greenhouse gases may have been added to an atmosphere by an asteroid impact, volcanic activity, or even massive forest fires.

Whatever the cause, during this warming episode, temperatures rose drastically. Much of the Earth became desert and over 90% of all species living at that time went extinct.

This is a good example of what can happen if our planet's essential cycles experience a big change.

Another important variable effected by the carbon cycle is the acidity of the ocean. Carbon dioxide can react with ocean water to form carbonic acid. This has been an important stabilizing force of the carbon cycle over the years, since the chemical equilibrium between carbon dioxide and carbonic acid means that the ocean can absorb or release carbon dioxide as atmospheric levels rise and fall.

However, as you might guess, increasing ocean acidity can mean trouble for sea life – and this might eventually pose a problem for other parts of the carbon system. Many forms of sea life that have shells, for example, can take carbon out of the water to create the calcium carbonate that they make their shells out of. If these species suffer, the ocean may lose some of its ability to remove carbon from the atmosphere.

Lastly, of course, there is the role of living things in the carbon cycle. The activity of plants and animals has been one of the major forces affecting changes to the carbon cycle in the past several billion years. Photosynthesizers have changed Earth's atmosphere and climate drastically by taking huge amounts of carbon out of the atmosphere and turning that carbon into cellular materials.

Those activities created free oxygen and the ozone layer, and generally set the stage for the evolution of animals that obtain their energy by breaking down the organic materials created by photosynthesizers and extracting the energy that the photosynthesizers used to make those molecules.

With one particular species of animals – humans – making big changes, the future of the Earth's carbon cycle is uncertain.

All such cycles in closed systems eventually correct themselves – but sometimes this happens through drastic population reduction of the offending species through starvation.

Steps of Carbon Cycle

Carbon in the Atmosphere

To become part of the carbon cycle, carbon atoms start out in gaseous form. Carbon dioxide gas – CO_2 – can be produced by inorganic processes, or by the metabolisms of living things.

Before Earth had life on it, carbon dioxide gas likely came from volcanic activity and asteroid impacts.

Today, carbon is also released into the atmosphere through the activities of living things, such the exhalations of animals, the actions of decomposer organisms, and the burning of wood and fossil fuels by humans.

However carbon dioxide gets into the atmosphere, CO_2 gas is the starting point of the carbon cycle. The next step is

Producers Absorb Carbon

"Producers" – organisms that produce food from sunlight, such as plants – absorb carbon dioxide from the atmosphere and use it to build sugars, lipids, proteins, and other essential building blocks of life.

For plants, CO_2 is absorbed through pores in their leaves called "stomata." Carbon dioxide enters the plant through the stomata and is incorporated into containing carbon compounds with the help of energy from sunlight.

Plants and other producer organisms such as cyanobacteria are crucial to life on Earth, because they can turn atmospheric carbon into living matter.

Producers are Eaten

"Consumers" are organisms that eat other living things. Animals are the most visible type of consumer in our ecosystems, though many types of microbes also fall into this category.

Consumers incorporate the carbon compounds from plants and other food sources when they eat them.

They use some of these carbon compounds from food to build their own bodies – but much of the food they eat is broken down to release energy, in a process that is almost the reverse of what producers do.

While producers use energy from sunlight to make bonds between carbon atoms – animals break these bonds to release the energy they contain, ultimately turning sugars, lipids, and other carbon compounds into single-carbon units. These are ultimately released into the atmosphere in the form of CO_2.

This process of "cellular respiration" – where oxygen gas is inhaled and carbon dioxide is exhaled – is a major source of carbon release back into the atmosphere.

But it's not always the last step of the carbon cycle. What about the carbon compounds that don't get eaten, or broken down by animals?

Decomposers Release Carbon

Plants and animals that die without being eaten by other animals are broken down by other organisms, called "decomposers."

Decomposers include many bacteria and some fungi. They usually only break down matter that is already dead, rather than catching and eating a living animal or plant.

Just like animals, decomposers break down the chemical bonds in their food molecules. They create many chemical products, including in some cases CO_2.

Carbon that isn't released back into the atmosphere in this way can also be released by.

Human Combustion

Humans are the only animals we know of who can create fire on purpose. And we set fire to things a lot.

Our cars are driven by burning fossil fuels – oil and gasoline, which are made of dead plant and animal material that spent millions of years buried deep in the Earth.

Many of our electrical power plants are powered by burning fossil fuels as well, including coal, which is another form of dead plant matter that was buried underground and transformed by geologic heat.

Lastly, humans also burn a lot of wood. We no longer burn wood to power our machines as we did in the 19th century, but now we often burn forests in order to clear land for agriculture, mining, and other purposes. About half of Earth's forests have been burned or otherwise destroyed by human activity to date.

The scientific community has raised alarms that by making significant changes to the Earth's carbon cycle, we may end up changing our climate or other important aspects of the ecosystem we rely upon to survive.

As a result, many scientists advocate to decrease the amount of carbon burned by humans by reducing car trips and electricity consumption, and investing in non-burning sources of energy such as solar power and wind power.

Examples of the Carbon Cycle

The carbon cycle consists of many parallel systems which can either absorb or release carbon. Together, these systems work to keep Earth's carbon cycle – and subsequently its climate and biosphere – relatively stable.

Here are some examples of parts of Earth's ecosystems that can absorb carbon, turn carbon into living matter, or release carbon back into the atmosphere.

Atmosphere

One major repository of carbon is the carbon dioxide in the Earth's atmosphere. Carbon forms a stable, gaseous molecule in combination with two atoms of oxygen.

In nature, this gas is released by volcanic activity, and by the respiration of animals who affix carbon molecules from the food they eat to molecules of oxygen before exhaling it.

Humans also release carbon dioxide into the atmosphere by burning organic matter such as wood and fossil fuels.

Carbon dioxide can be removed from the atmosphere by plants, which take the atmospheric carbon and turn it into sugars, proteins, lipids, and other essential molecules for life.

It can also be removed from the atmosphere by absorption into the ocean, whose water molecules can bond with carbon dioxide to form carbonic acid.

In recent years, scientists have raised concerns that by cutting down about half of Earth's forests, humans may be decreasing the Earth's ability to remove carbon dioxide from the atmosphere at the same time they're adding new sources by burning wood and fossil fuels.

Many organizations that hope to fight man-made climate change now plant trees which can remove carbon dioxide from the atmosphere, as well as advocating for alternative energy sources and less burning of fossil fuels such as gasoline, oil, and coal.

Lithosphere

The Earth's crust – called the "lithosphere" from, "Litho" for "stone" and "sphere" for globe – can also release carbon dioxide into Earth's atmosphere. This gas can be created by chemical reactions in the Earth's crust and mantel.

Volcanic activity can result in natural releases of carbon dioxide. Some scientists believe that widespread volcanic activity may be to blame for the warming of the Earth that caused the Permian extinction.

While the Earth's crust can add carbon to the atmosphere, it can also remove it. Movements of the Earth's crust can bury carbon-containing chemicals such as dead plants and animals deep underground, where their carbon cannot escape back into the atmosphere.

Over millions of years, these underground reservoirs of organic matter liquefy and become coal, oil, and gasoline. In recent years, humans have begun releasing much of this sequestered carbon back into the atmosphere by burning these materials to power cars, power plants, and other human equipment.

Biosphere

Among living things, some remove carbon from the atmosphere, while others release it back. The most noticeable participants in this system are plants and animals.

Plants remove carbon from the atmosphere. They don't do this as a charitable act; atmospheric carbon is actually the "food" which plants use to make sugars, proteins, lipids, and other essential molecules for life.

Plants use the energy of sunlight, harvested through photosynthesis, to build these organic compounds out of carbon dioxide and other trace elements. Indeed, the term "photosynthesis" comes from, "photo" for "light" and "synthesis" for "to put together".

Here lies arguably the most important part of the carbon cycle: all life is made of carbon. Without plants or other organisms that could turn inorganic carbon compounds like carbon dioxide into organic compounds, life could not exist.

Indeed, none of the building materials for our cells, from our DNA to our cell membranes, could exist without this ability of photosynthesizers to turn carbon dioxide into life.

In a gracefully balanced set of chemical reactions, animals eat plants (and other animals), and take these synthesized molecules apart again. Animals get their fuel from the chemical energy plants have stored in the bonds between carbon atoms and other atoms during photosynthesis.

In order to do that, animal cells dissemble complex molecules such as sugars, fats, and proteins all the way down to single-carbon units – molecules of carbon dioxide, which are produced by reacting carbon-containing food molecules with oxygen from the air.

In this way, most of the carbon eaten by animals ends up back where it started before it was absorbed by a plant – as part of a carbon dioxide molecule in the atmosphere.

Carbon in plants and animals that is not consumed by other animals can be broken down by other living things until it becomes carbon dioxide again, or can be sequestered deep in the Earth as fossil fuels.

Oceans

The Earth's oceans have the ability to both absorb and release carbon dioxide. When carbon dioxide from the atmosphere comes into contact with ocean water, it can react with the water molecules to form carbonic acid – a dissolved liquid form of carbon.

Like most chemical reactions, the rate of this reaction is determined by the equilibrium between the products and the reactants.

When there is more carbonic acid in the ocean compared to carbon dioxide in the atmosphere, some carbonic acid may be released into the atmosphere as carbon dioxide.

On the other hand, when there is more carbon dioxide in the atmosphere, more carbon dioxide will be converted to carbonic acid, and ocean acidity levels will rise.

Some scientists have raised concerns that acidity is rising in some parts of the ocean, possibly as a result of increased carbon dioxide in the atmosphere due to human activity.

Although these changes in ocean acidity may sound small by human standards, many types of sea life depend on chemical reactions that need a highly specific acidity level to survive.

As a result, there is concern that increasing ocean acidity due to carbonic acid may contribute to the die-offs of some marine ecosystems, and even to extinctions of marine species.

Human Impact on Carbon Cycle

Burning of Fossil Fuels

When oil or coal is burned, carbon is released into the atmosphere at a faster rate than it is removed. As a result, the concentration of carbon dioxide in the atmosphere increases. Natural gas, oil and coal are fossil fuels that are commonly burned to generate electricity in power plants, for transportation, in homes and in other industrial complexes. The primary industrial activities that emit carbon dioxide and affect the carbon cycle are petroleum refining, paper, food and mineral production, mining and the production of chemicals.

Carbon Sequestration

When plants remove carbon dioxide from the air and store it, the process is called carbon sequestration. Agricultural and forestry methods can affect how much carbon dioxide is removed from the atmosphere and stored by the plants. These sinks of carbon dioxide can be farms, grasslands or forests. Human activity in managing farmland or forests affects the amount of carbon dioxide removed from the atmosphere by plants and trees. These sinks of carbon dioxide affect the carbon cycle by decreasing the amount of carbon dioxide in the air.

Deforestation

Deforestation is the permanent removal of trees from forests. Permanent removal of the trees means new trees will not be replanted. This large-scale removal of trees from forests by people results in increased levels of carbon dioxide in the atmosphere because trees are no longer absorbing carbon dioxide for photosynthesis. As a result, the carbon cycle is affected. According to National Geographic, agriculture is the primary cause of deforestation. Farmers remove trees on a large-scale basis to increase acreage for crops and livestock.

Geologic Sequestration

Human activity can affect the carbon cycle by capturing carbon dioxide and storing it underground rather than permitting it to be released into the atmosphere. This process is called geologic sequestration. According to the U.S. Environmental Protection Agency, geologic sequestration could retain large quantities of carbon dioxide for extended periods of time and consequently reduce the concentrations of carbon dioxide above-ground.

Nitrogen Cycle

Nitrogen cycle is the circulation of nitrogen in various forms through nature. Nitrogen, a component of proteins and nucleic acids, is essential to life on Earth. Although 78 percent by volume of the atmosphere is nitrogen gas, this abundant reservoir exists in a form unusable by most organisms. Through a series of microbial transformations, however, nitrogen is made available to plants, which in turn ultimately sustain all animal life. The steps, which are not altogether sequential, fall into the following classifications: nitrogen fixation, nitrogen assimilation, ammonification, nitrification, and denitrification.

Nitrogen fixation, in which nitrogen gas is converted into inorganic nitrogen compounds, is mostly (90 percent) accomplished by certain bacteria and blue-green algae. A much smaller amount of free nitrogen is fixed by abiotic means (e.g., lightning, ultraviolet radiation, electrical equipment) and by conversion to ammonia through the Haber-Bosch process.

Nitrates and ammonia resulting from nitrogen fixation are assimilated into the specific tissue compounds of algae and higher plants. Animals then ingest these algae and plants, converting them into their own body compounds.

The remains of all living things—and their waste products—are decomposed by microorganisms in the process of ammonification, which yields ammonia (NH_3) and ammonium (NH_4+). (Under anaerobic, or oxygen-free, conditions, foul-smelling putrefactive products may appear, but they too are converted to ammonia in time.) Ammonia can leave the soil or be converted into other nitrogen compounds, depending in part on soil conditions.

Nitrification, a process carried out by nitrifying bacteria, transforms soil ammonia into nitrates (NO_3-), which plants can incorporate into their own tissues.

Nitrates also are metabolized by denitrifying bacteria, which are especially active in water-logged anaerobic soils. The action of these bacteria tends to deplete soil nitrates, forming free atmospheric nitrogen.

Steps of Nitrogen Cycle

Nitrogen Fixation

Nitrogen fixation is the process by which gaseous nitrogen (N_2) is converted to ammonia (NH_3 or NH_4^+) via biological fixation or nitrate (NO_3^-) through high-energy physical processes. N_2 is extremely stable and a great deal of energy is required to break the bonds that join the two N atoms. N_2 can be converted directly into NO_3^- through processes that exert a tremendous amount of heat, pressure, and energy. Such processes include combustion, volcanic action, lightning discharges, and industrial means. However, a greater amount of biologically available nitrogen is naturally generated via the biological conversion of N_2 to NH_3/ NH_4^+. A small group of bacteria and cyanobacteria are capable using the enzyme nitrogenase to break the bonds among the molecular nitrogen and combine it with hydrogen.

Nitrogenase only functions in the absence of oxygen. The exclusion of oxygen is accomplished by many means. Some bacteria live beneath layers of oxygen-excluding slime on the roots of certain plants. The most important soil dwelling bacteria, Rhizobium, live in oxygen-free zones in nodules on the roots of legumes and some other woody plants. Aquatic filamentous *cyanobacteria* utilize oxygen-excluding cells called *heterocysts*.

Nitrification

Nitrification is a two-step process in which NH_3/ NH_4^+ is converted to NO_3^-. First, the soil bacteria *Nitrosomonas* and *Nitrococcus* convert NH_3 to NO_2^-, and then another soil bacterium, *Nitrobacter*, oxidizes NO_2^- to NO_3^-. These bacteria gain energy through these conversions, both of which require oxygen to occur.

Assimilation

Assimilation is the process by which plants and animals incorporate the NO_3^- and ammonia formed through nitrogen fixation and nitrification. Plants take up these forms of nitrogen through their roots, and incorporate them into plant proteins and nucleic acids. Animals are then able to utilize nitrogen from the plant tissues.

Ammonification

Assimilation produces large quantities of organic nitrogen, including proteins, amino acids, and nucleic acids. *Ammonification* is the conversion of organic nitrogen into ammonia. The ammonia produced by this process is excreted into the environment and is then available for either nitrification or assimilation.

Denitrification

Denitrification is the reduction of NO_3^- to gaseous N_2 by anaerobic bacteria. This process only occurs where there is little to no oxygen, such as deep in the soil near the water table. Hence, areas such as wetlands provide a valuable place for reducing excess nitrogen levels via denitrification processes.

Nitrogen Fixation

A relatively small amount of ammonia is produced by lightning. Some ammonia also is produced industrially by the Haber-Bosch process, using an iron-based catalyst, very high pressures and fairly high temperature. But the major conversion of N2 into ammonia, and thence into proteins, is achieved by microorganisms in the process called nitrogen fixation (or dinitrogen fixation).

The table below shows some estimates of the amount of nitrogen fixed on a global scale. The total biological nitrogen fixation is estimated to be twice as much as the total nitrogen fixation by non-biological processes.

Type of fixation	N_2 fixed (10^{12} g per year, or 10^6 metric tons per year)
Non-biological	
Industrial	about 50
Combustion	about 20
Lightning	about 10
Total	about 80

Biological	
Agricultural land	about 90
Forest and non-agricultural land	about 50
Sea	about 35
Total	about 175

To illustrate the importance of biological nitrogen fixation, the image below shows part of the Lower Sonoran desert in Arizona. Every plant that we see in this scene depends ultimately on biological nitrogen fixation. Both free-living cyanobacteria and the cyanobacterial associates of lichens initially contributed nitrogen to the soil by forming a cryptobiotic crust. Now numerous leguminous plants occur in this desert, with nitrogen-fixing Rhizobium in their root nodules. Examples are the green-stemmed brush-like trees at the right and left of the image (Parkinsonia species, common name "paloverde"), and several acacias and mesquites

Mechanism of biological nitrogen fixation

Biological nitrogen fixation can be represented by the following equation, in which two moles of ammonia are produced from one mole of nitrogen gas, at the expense of 16 moles of ATP and a supply of electrons and protons (hydrogen ions):

$$N_2 + 8H+ + 8e^- + 16\ ATP = 2NH_3 + H_2 + 16ADP + 16\ Pi$$

This reaction is performed exclusively by prokaryotes (the bacteria and related organisms), using an enzyme complex termed nitrogenase. This enzyme consists of two proteins - an iron protein and a molybdenum-iron protein, as shown below.

The reactions occur while N_2 is bound to the nitrogenase enzyme complex. The Fe protein is first reduced by electrons donated by ferredoxin. Then the reduced Fe protein binds ATP and reduces the molybdenum-iron protein, which donates electrons to N_2, producing HN=NH. In two further cycles of this process (each requiring electrons donated by ferredoxin) HN=NH is reduced to H_2N-NH_2, and this in turn is reduced to $2NH_3$.

Depending on the type of microorganism, the reduced ferredoxin which supplies electrons for this process is generated by photosynthesis, respiration or fermentation.

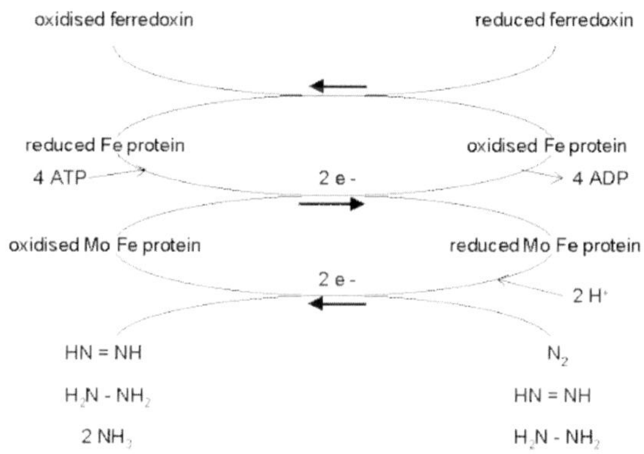

There is a remarkable degree of functional conservation between the nitrogenase proteins of all nitrogen-fixing bacteria. The Fe protein and the Mo-Fe protein have been isolated from many of these bacteria, and nitrogen fixation can be shown to occur in cell-free systems in a laboratory when the Fe protein of one species is mixed with the Mo-Fe protein of another bacterium, even if the species are very distantly related.

The Nitrogen-fixing Organisms

All the nitrogen-fixing organisms are prokaryotes (bacteria). Some of them live independently of other organisms - the so-called free-living nitrogen-fixing bacteria. Others live in intimate symbiotic associations with plants or with other organisms (e.g. protozoa). Examples are shown in the table below.

Examples of nitrogen-fixing bacteria (* denotes a photosynthetic bacterium)			
Free living		Symbiotic with plants	
Aerobic	Anaerobic	Legumes	Other plants
Azotobacter Beijerinckia Klebsiella (some) Cyanobacteria (some)*	Clostridium (some) Desulfovibrio Purple sulphur bacteria* Purple non-sulphur bacteria* Green sulphur bacteria*	Rhizobium	Frankia Azospirillum

A point of special interest is that the nitrogenase enzyme complex is highly sensitive to oxygen. It is inactivated when exposed to oxygen, because this reacts with the iron component of the proteins. Although this is not a problem for anaerobic bacteria, it could be a major problem for the aerobic species such as cyanobacteria (which generate oxygen during photosynthesis) and the free-living aerobic bacteria of soils, such as Azotobacter and Beijerinckia. These organisms have various methods to overcome the problem. For example, Azotobacter species have the highest known rate of respiratory metabolism of any organism, so they might protect the enzyme by maintaining a very low level of oxygen in their cells. Azotobacter species also produce copious amounts of extra-cellular polysaccharide. By maintaining water within the polysaccharide slime layer, these bacteria can limit the diffusion rate of oxygen to the cells. In the symbiotic nitrogen-fixing organisms such as Rhizobium, the root nodules can contain oxygen-scavenging molecules such as leghaemoglobin, which shows as a pink colour when the active nitrogen-fixing nodules of legume roots are cut open. Leg haemoglobin may regulate the supply of oxygen to the nodule tissues in the same way as haemoglobin regulates the supply of oxygen to mammalian tissues. Some of the cyanobacteria have yet another mechanism for protecting nitrogenase: nitrogen fixation occurs in special cells (heterocysts) which possess only photosystem I (used to generate ATP by light-mediated reactions) whereas the other cells have both photosystem I and photosystem II (which generates oxygen when light energy is used to split water to supply H_2 for synthesis of organic compounds).

Symbiotic Nitrogen Fixation

Legume Symbioses

The most familiar examples of nitrogen-fixing symbioses are the root nodules of legumes (peas, beans, clover, etc.).

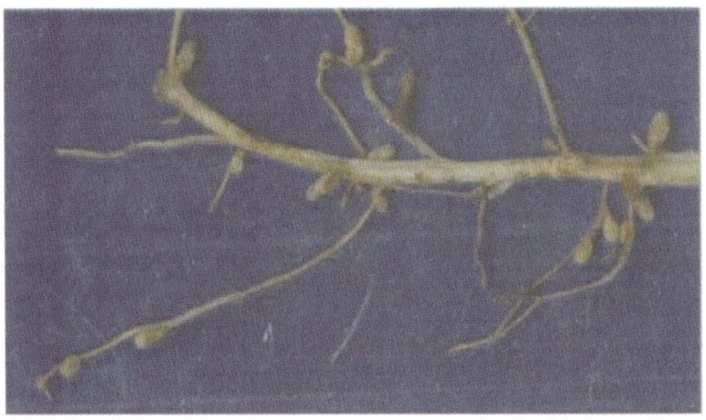

Part of a clover root system bearing naturally occurring nodules of Rhizobium. Each nodule is about 2-3 mm long.

Clover root nodules at higher magnification, showing two partly crushed nodules (arrowheads) with pink-coloured contents. This colour is caused by the presence of the pigment leghaemoglobin - a unique metabolite of this type of symbiosis. Leghaemoglobin is found only in the nodules and is not produced by either the bacterium or the plant when grown alone.

In these leguminous associations the bacteria usually are Rhizobium species, but the root nodules of soybeans, chickpea and some other legumes are formed by small-celled rhizobia termed Bradyrhizobium. Nodules on some tropical leguminous plants are formed by yet other genera. In all cases the bacteria "invade" the plant and cause the formation of a nodule by inducing localised proliferation of the plant host cells. Yet the bacteria always remain separated from the host cytoplasm by being enclosed in a membrane - a necessary feature in symbioses.

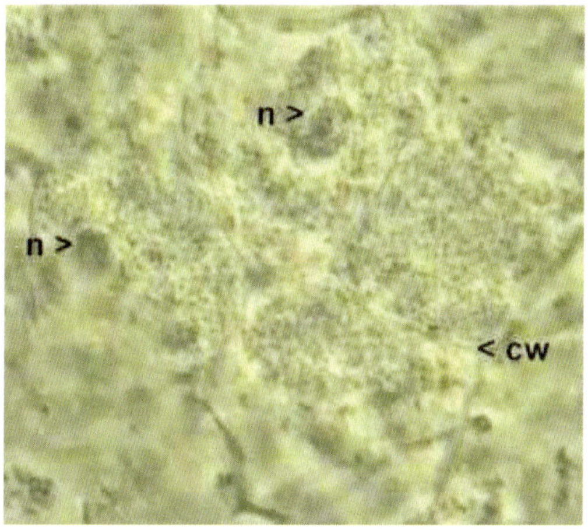

Part of a crushed root nodule of a pea plant, showing four root cells containing colonies of Rhizobium.

The nuclei (n) of two root cells are shown; cw indicates the cell wall that separates two plant cells. Although it cannot be seen clearly in this image, the bacteria occur in clusters which are enclosed in membranes, separating them from the cytoplasm of the plant cells.

In nodules where nitrogen-fixation is occurring, the plant tissues contain the oxygen-scavenging molecule, leghaemoglobin (serving the same function as the oxygen-carrying haemoglobin in blood). The function of this molecule in nodules is to reduce the amount of free oxygen, and thereby to protect the nitrogen-fixing enzyme nitrogenase, which is irreversibly inactivated by oxygen

Associations with Frankia

Frankia is a genus of the bacterial group termed actinomycetes - filamentous bacteria that are noted for their production of air-borne spores. Included in this group are the common soil-dwelling Streptomyces species which produce many of the antibiotics used in medicine . Frankia species are slow-growing in culture, and require specialised media, suggesting that they are specialised symbionts. They form nitrogen-fixing root nodules (sometimes called actinorhizae) with several woody plants of different families, such as alder (Alnus species), sea buckthorn (Hippophae rhamnoides, which is common in sand-dune environments) and Casuarina (a Mediterranean tree genus). Figure A (below) shows a young alder tree (Alnus glutinosa) growing in a plant pot, and Figure B shows part of the root system of this tree, bearing the orange-yellow coloured nodules (arrowheads) containing Frankia.

Alder and the other woody hosts of Frankia are typical pioneer species that invade nutrient-poor soils. These plants probably benefit from the nitrogen-fixing association, while supplying the bacterial symbiont with photosynthetic products.

Cyanobacterial Associations

The photosynthetic cyanobacteria often live as free-living organisms in pioneer habitats such as desert soils or as symbionts with lichens in other pioneer habitats. They also form symbiotic associations with other organisms such as the water fern *Azolla*, and cycads.The association with *Azolla*, where cyanobacteria (*Anabaena azollae*) are harboured in the leaves, has sometimes been shown to be important for nitrogen inputs in rice paddies, especially if the fern is allowed to grow and then ploughed into the soil to release nitrogen before the rice crop is sown. A symbiotic association of cyanobacteria with cycads is shown below. The first image shows a pot-grown plant. The second image shows a close-up of the soil surface in this pot. Short, club-shaped, branching roots have grown into the aerial environment. These aerial roots contain a nitrogen-fixing cyanobacterial symbiont.

In addition to these intimate and specialised symbiotic associations, there are several free-living nitrogen-fixing bacteria that grow in close association with plants. For example, Azospirillum species have been shown to fix nitrogen when growing in the root zone (rhizosphere) or tropical grasses, and even of maize plants in field conditions. Similarly, Azotobacter species can fix nitrogen in the rhizosphere of several plants. In both cases the bacteria grow at the expense of sugars and other nutrients that leak from the roots. However, these bacteria can make only a small contribution to the nitrogen nutrition of the plant, because nitrogen-fixation is an energy-expensive process, and large amounts of organic nutrients are not continuously available to microbes in the rhizosphere.

This limitation may not apply to the bacteria that live in root nodules or other intimate symbiotic associations with plants. It has been estimated that nitrogen fixation in the nodules of clover roots or other leguminous plants may consume as much as 20% of the total photosynthate.

Nitrification

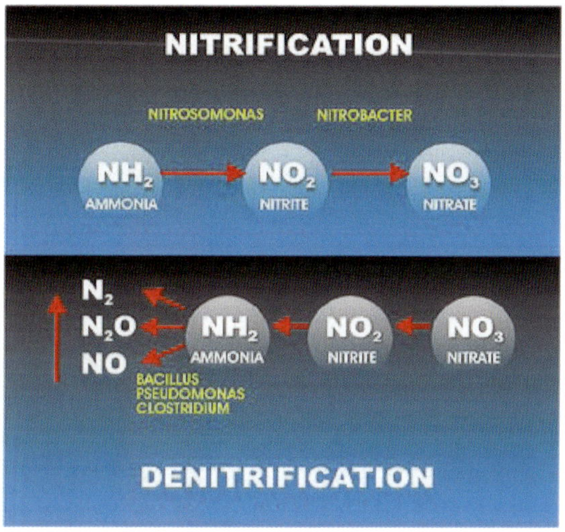

Nitrification involves the conversion of ammonia into nitrite by the Nitrosomonas bacteria. Nitrite is then converted to nitrate by Nitrobacter. Simultaneously, denitrification occurs wherein nitrate is converted to nitrite then to ammonia which when acted upon by Bacillus, Psuedomonas and Clostridium, is converted into nitrogen gas, nitrous oxide and nitric oxide, all nontoxic, and are released in the process.

When fish and prawns are cultured and fed protein-rich feeds, uneaten food particles and by-products like amino acids and proteins, become major sources of organic compounds that accumulate on the pond bottom. These wastes are ultimately metabolized by bacteria to ammonia. Bacteria in NS Series SPO use dissolved organic sources of carbon from proteins, fats, and carbohydrates to build their body components. Conversion of dissolved organic material by these bacteria is a precursor to nitrification. The wastes excreted by fish are in the form of ammonia and urea. Their amounts accelerates as stocking and feeding rates increase. Most of the ammonia is taken up directly by the algae. When nitrate is used, it must first be reduced to ammonia (the reverse of nitrification). However, most of the ammonia will be converted by nitrifying bacteria to nitrite, and then nitrate.

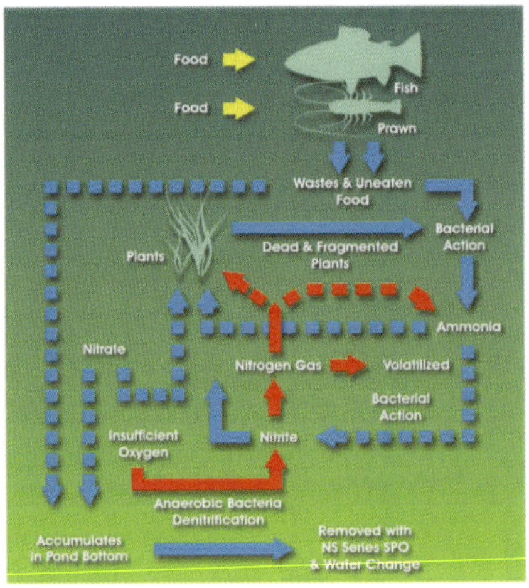

Excess ammonia is toxic to fish and must be removed or converted into non-toxic substances. It exists as a mixture of two forms, unionized ammonia (NH_3) and ionized ammonia (NH_4+). They are converted from one to the other at an equal rate depending upon the pH and temperature of the water. Ammonia, consumed by the bacteria in NS Series SPO, is initially converted it into nitrite and subsequently into nitrate. Nitrite (NO_2) is toxic to fish and can reduce growth and adversely affect fish health with high mortalities. Nitrate (NO_3) is the end product of nitrification and is normally not toxic to fish even in high concentrations. Nitrate is tolerated by some fish species however, ideally it should be maintained below 20 ppm. Nitrate accumulation can be controlled by regular water changes and can also be converted to nitrogen gas.

In properly balanced aquaculture environment, ammonia and nitrite levels should always be zero. Their presence is an indication that something in the system is out of balance and s an indication for you to start corrective action. Even If low levels of these persist in the aquaculture environment, they will cause problems for your prawns and fishes, resulting in increased susceptibility to other diseases.

When oxygen is absent, denitrification can occur when certain anaerobic bacteria obtain their oxygen by removing it from the nitrate (NO_3-) ion, finally leaving nitrogen gas or organic nitrogen compounds. Nitrogen gas is then either volatilized to the atmosphere or converted by biological fixation to ammonia. NS Series SPO can reduce nitrate. Nitrate is used as a nitrogen source by reducing it and incorporating the nitrogen atoms into organic molecules. Denitrification generally requires anoxic conditions and adequate soluble organic carbon. This combination of conditions is usually lacking in many aquaculture environments and this prevents the complete removal of nitrate.

Assimilation

There are different sources of nitrogen from which plant can directly or indirectly in take or assimilate nitrogen.

a) Atmospheric nitrogen

   Atmospheric gases contain nitrogen in form of nitrogen gas and oxide of nitrogen majority of plants cannot assimilate it directly. But some bacteria, blue green algae and leguminous plants can fix the atmospheric nitrogen into available form and assimilate.

b) Nitrate, nitrites, ammonia in soil (Inorganic nitrogen)

Is soil nitrogen present in from of ammonium salts. Well nitrates, nitrires and etc as inorganic nitrogen. Out of these nitrate form can be easily assimilated by higher plants.

c) Amino acids in soil (organic from)

Many soil micro organisms utilize nitrogen in amino acid form which available in soil by decomposition of dead bodies.

d) Organic nitrogenous compounds (Proteins & amino acids, nucleic Acids) in bodies of insects

Insectivorous plants fulfill their nitrogen requirement by trapping and digesting insects. These plants include Picher plants, (nepenthes), sundew plant (drosera), Bladderwort (Utricularia), Venus fly trap (Diones) etc.

Mechanism of Nitrogen Assimilation in Plants

The process by which nitrogen is taken up by plants for the synthesis of proteins is called nitrogen assimilation. Higher plant absorbs nitrogen preferably in form nitrate.

Therefore nitrogen at whatever form it is present should be transformed in to available nitrate form. Then the nitrates after entering into the root converted or reduced into among and utilized in the tissues making various organic compounds.

Nitrogen Assimilation in Plants Takes Place through following steps:

I. Nitrogen fixation

First atmospheric nitrogen either non-symbiotically or symbiotically converted into nitrates and makes it available in soil or root nodules. Some tree living bacteria like Azatobacter, clostridium, chlorobium and some symbiotic bacteria like rhizobium blue green algae like nostoc help in nitrogen fixation.

II. Nitrate reduction

After fixation in to nitrates, these are enzymatic-ally reduced to nitrites and ammonia and then can be assimilated. Stepwise reduction of nitrate takes place by addition of two electrons and ultimately reduced to ammonia. The electrons supplied to this system by reduces co-enzyme NADH and NADPH.

(a) Reduction of nitrate to nitrite

Reduction of nitrate to nitrite takes place in presence of the enzyme nitrate reductase which require reduced co-enzyme NADH or NSDPH for its activation.

This enzyme which is molyboflavo protein with an operative syllahydryl group was first isolated by Evans and nason (1953) from neuro-spora and soyabean leaves. The enzyme isolated from beurospora could utilize only reduced co-enzyme isolated from beurospora could utilize only reduced co-enzyme NADPH 2while the enzyme isolated from soyabean leaves could utilize both the reduced coenzyme NADH and NADPH.

The enzyme contain FAD as its prosthetic group with which is associated molybdenum actual reduction of takes place as shown.

Electron are transferred from reduced coenzyme to FAD which become reduced FADH2 from reduced FADH2 the electrons are finally transferred to NO3 through molybdenum so that NO3 and H2O are formed.

This nitrate reduction takes place chiefly in green leaves and roots.

(b) Reduction of nitrite to ammonia:

It takes place in the presence the enzyme nitrite re-educates which requires NADH and NADH2. This reaction occurs in leaves in the presence of light more rapidly and in dark at slow rate.

Manganese seems to be associated with the prosthetic group of this enzyme which was first isolated by Nason et al from Neurospore and soyabean leaves.

Previously it was thought that the reduction of nitrite to ammonia involved the formation of two intermediate compounds like hyponitrites and hydroxylamine but it is uncertain due to:

1. Hyponitrite is quite unstable
2. Hydroxylamine is toxic
3. Moreover, they have never been observed in Free State in the cells.

Moreover it is believed now that, hyponitrite and hydroxylamine may at the best formed at the surface of enzymes and readily leave the surface of enzymes when they completely reduced in to ammonia or other intermediates.

After ammonia is formed this ammonia enters into the biosynthetic path ways of plant cell, like reductive amination, transpiration etc to produce different amino acids. Usually ammonia reacts with the intermediates of kreb's cycle in presence of NADPH2. Amino-acids in cells take part in the synthesis of protein and other nitrogenous compounds that help in body building.

Ammonification

Ammonification, in chemistry, is defined as the saturation with ammonia or any one of its compounds. Strictly speaking, ammonification refers to any chemical reaction that generates ammonia (NH3) as an end product (or its ionic form, ammonium, NH4+). Ammonification can occur through various inorganic reactions or due to the metabolic functions of microorganisms, plants, and animals. In the ecological context, however, ammonification refers to the processes by which organically bound forms of nitrogen occurring in dead biomass (such as amino acids and proteins) are oxidized into ammonia and ammonium. The ecological process of ammonification is carried out in soil and water by a great diversity of microbes and is one of the many types of chemical transformations that occur during the decomposition of dead organic matter.

Ammonification is a key component in the nitrogen cycle of ecosystems. The nitrogen cycle consists of a complex of integrated processes by which nitrogen circulates among its major compartments in the atmosphere, water, soil, and organisms. During various phases of the nitrogen cycle, this element is transformed among its various organic and inorganic compounds.

As with all components of the nitrogen cycle, the proper functioning of ammonification is critical to the health of ecosystems. In the absence of ammonification, organic forms of nitrogen would accumulate in large quantities. Because growing plants need access to inorganic forms of nitrogen, particularly ammonium and nitrate (NO3–), the oxidation of organic nitrogen of dead biomass through ammonification is necessary for maintenance of the productivity of species and ecosystems.

Nitrogen is one of the most abundant elements in the tissues of all organisms and is a component of many biochemicals, particularly amino acids, proteins, and nucleic acids. Consequently, nitrogen is one of the critically important nutrients and is required in relatively large quantities by all organisms. Animals receive their supply of nitrogen through the foods they eat, but plants must assimilate inorganic forms of this nutrient from their environment.

However, the rate at which the environment can supply inorganic nitrogen is limited and usually small in relation to the metabolic demands of plants. Therefore, the availability of inorganic forms of nitrogen is frequently a limiting factor for the productivity of plants. This is a particularly common occurrence for plants growing in terrestrial and marine environments and, to a lesser degree, in fresh waters (where phosphate supply is usually the primary limiting nutrient, followed by nitrate).

The dead biomass of plants, animals, and microorganisms contains large concentrations of organically bound nitrogen in various forms, such as proteins and amino acids. The process of decomposition is responsible for recycling the inorganic constituents of the dead biomass and preventing it from accumulating in large unusable quantities. Decomposition is, of course, mostly carried out through the metabolic functions of a diverse array of bacteria, fungi, actinomycetes, other microorganisms, and some animals. Ammonification is a particular aspect of the more complex process of organic decay, specifically referring to the microbial conversion of organic-nitrogen into ammonia (NH_3) or ammonium (NH_4+).

Ammonification occurs under oxidizing conditions in virtually all ecosystems and is carried out by virtually all microorganisms that are involved in the decay of dead organic matter. In situations where oxygen is not present, a condition referred to as anaerobic, different microbial decay reactions occur; these produce nitrogen compounds known as amines.

The microbes derive some metabolically useful energy from the oxidation of organic-nitrogen to ammonium. In addition, much of the ammonium is assimilated and used as a nutrient for the metabolic purposes of the microbes. However, if the microbes produce ammonium in quantities that exceed their own requirements, as is usually the case, the surplus is excreted into the ambient environment (such as the soil), and is available for use as a nutrient by plants, or as a substrate for another microbial process, known as nitrification. Animals, in contrast, mostly excrete urea or uric acid in their nitrogen-containing liquid wastes (such as urine), along with diverse organic-nitrogen compounds in their feces. The urea, uric acid, and organic nitrogen of feces are all substrates for microbial ammonification.

One of the most elementary of the ammonification reactions is the oxidation of the simple organic compound urea (CO (NH2)2) to ammonia through the action of a microbial enzyme known as urease. (Note that two units of ammonia are produced for every unit of urea that is oxidized.) Urea is a commonly utilized agricultural fertilizer, used to supply ammonia or ammonium for direct uptake by plants, or as a substrate for the microbial production of nitrate through nitrification.

Ammonium is a suitable source of nitrogen uptake for many species of plants, particularly those that live in acidic soils and waters. However, most plants that occur in non-acidic soils cannot utilize ammonium very efficiently, and they require the anion nitrate (NO3+) as their source of nitrogen uptake. The nitrate is generally derived by the bacterial oxidation of ammonium to nitrite, and then to nitrate, in an important ecological process known as nitrification. Because the species of bacteria that carry out nitrification are extremely intolerant of acidity, this process does not occur at significant rates in acidic soils or waters. This is the reason why plants growing in acidic habitats can only rely on ammonium as their source of nitrogen nutrition.

Because ammonium is a positively charged cation, it is held relatively strongly by ion-exchange reactions occurring at the surfaces of clay minerals and organic matter in soils. Consequently, ammonium is not leached very effectively by water as it percolates downward through the soil. This is in contrast to nitrate, which is highly soluble in soil water and is leached readily. As a result, nitrate pollution can be an important problem in agricultural areas that have been heavily fertilized with nitrogen-containing fertilizers.

Humans have a major influence on the nitrogen cycle, especially through the use of fertilizers in agriculture. Under nutrient-limited conditions, farmers commonly attempt to increase the availability of soil nitrogen, particularly as nitrate, and to a lesser degree, as ammonium. Rates of fertilization in intensive agricultural systems can exceed 446.2 lb/ac (500 kg/ha) of nitrogen per year. The nitrogen in the fertilizer may be added as ammonium nitrate (NO4 NH4) or as urea. The latter compound must be ammonified before inorganic forms of nitrogen are present, that is, the ammonium and nitrate that can be taken up by plants. In some agricultural systems, compost or other organic materials may be added to soils as a conditioner and fertilizer. In such cases, the organic nitrogen is converted to available ammonium through microbial ammonification, and nitrate may subsequently be generated through nitrification.

In situations where the rates of fertilization are excessive, the ability of the ecosystem to assimilate the nitrogen input becomes satiated. Although the ammonium produced by ammonification does not leach readily, the nitrate does, and this can lead to the.

Denitrification

The natural cycle of denitrification comprises a cascade of different enzymes that stepwise reduce nitrate to dinitrogen:

$$NO_3 \rightarrow NO_2 \rightarrow NO \rightarrow N_2O \rightarrow N_2$$

Hence, denitrification corresponds to the part of the biological nitrogen cycle that is opposed to nitrogen fixation. The reduction of nitric oxide to nitrous oxide is mediated in nature by the NO reductases (NOR), which occur in bacteria (NorBC) as well as in fungi (P450nor). The principal reaction scheme of NO reduction corresponds to the equation:

$$2NO + 2H^+ + 2e^- \rightarrow N_2O + H_2O$$

Bacterial and fungal NORs are fundamentally different in the nature of their active sites. Correspondingly, very different reaction mechanisms have been proposed for these two classes of enzymes.

In the biological denitrification cycle, nitrous oxide is reduced to dinitrogen by the N2O reductases (N2OR) following the principal reaction scheme:

$$N_2O + 2H^+ + 2e^- \rightarrow N_2 + H_2O$$

The different classes of NO and N2O reductases are described in more detail in the following sections.

Bacterial NORs (NorBC)

These enzymes are closely related to the respiratory heme-copper oxydases, the so called Cytochrome c-Oxidases (CCO). Figure shows the active site of the NorBC from Pseudomonas aeruginosa in the diferric resting state, which consists of a heme b3 with axial (proximal) histidine coordination and the so called FeB, a non-heme iron centre. The latter one is coordinated by three histidines and a glutamate.

Bacterial respiratory NOR: NorBC

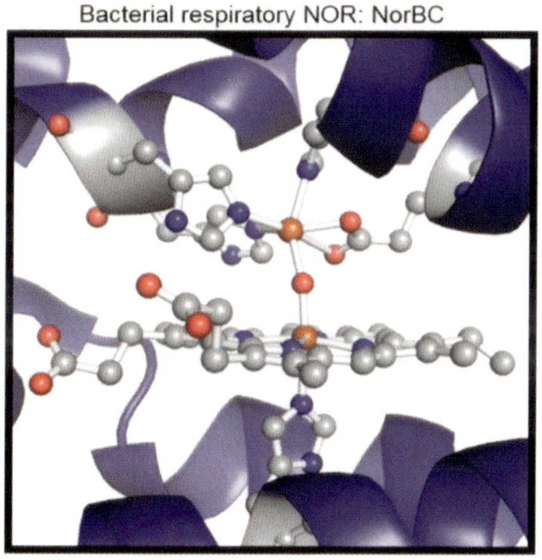

Figure: Oxo-bridged, differric resting state of the NorBC active site (PDB 3OoR)

Despite much effort, the molecular mechanism of NO reduction by NorBCs has remained largely elusive. Of the many possible mechanisms that have been proposed, the ones summarized in figure represent the most likely scenarios. Initial work by Girsch and de Vries favours the trans mechanism, where catalysis starts from the diferrous form of the enzyme, and each one of the iron centres binds one NO ligand first, followed by a radical-type N-N coupling mechanism. However, heme and non-heme ferrous NO complexes as invoked in this mechanism are typically very stable and unreactive, which poses the question of how these types of complexes could serve as reactive intermediates in the mechanism of cNORs. In terms of the heme, reactivity of the ferrous NO complex could be increased via the trans effect in a corresponding 6C system, although the radical reactivity of such complexes is still intrinsically low as evident from corresponding diferric hyponitrite (N2O22-) complexes, which have a propensity to decompose into the corresponding ferrous heme-nitrosyls,

$$\left[(L)Fe^{III}(porph)-N_2O_2-Fe^{III}(Porph)(L)\right] \rightarrow 2\left[Fe^{II}(Porph)(L)(NO)\right]$$

both in the absence and presence of an axial N-donor ligand L.

The cis-heme b3 mechanism in figure, middle is inspired by the mechanism of Cyt. P450nor, where one molecule of NO is bound first to the heme and subsequently activated for reaction with the second NO. A variation of this pathway where the heme-bound NO interacts with the FeB center is favoured by computational methods, according to Blomberg and co-workers. In this case, the reaction would also start from the diferrous form of the enzyme. The first NO binds to the heme, forming a 6C ferrous heme-nitrosyl complex that is activated by interaction with the FeB centre. The second NO then attacks the bound NO and, at the same time, coordinates with its O-atom at the non-heme iron centre. This leads to the direct generation of a bridged hyponitrite complex, whereas the non-heme iron-nitrosyl complex never forms. According to the latest study, this mechanism is energetically favourable compared to the trans mechanism, which is disfavoured by high activation energies.

The equivalent of the cis-heme b3 mechanism where the whole reaction proceeds at the non-heme iron center has also been proposed. In this case, the formation of a ferrous dinitrosyl complex has been proposed. However, no further evidence for this mechanism is available, and on the contrary, arguments can be made why this mechanism is unlikely. First, a corresponding ferrous dinitrosyl complex has never been observed in any enzyme or model system. Second, in analogous (but more reduced) DNICs, the Fe-NO units are strongly antiferromagnetically coupled, which forces the two NO ligands to have their unpaired electrons aligned with parallel spins. The necessary spin-flip of one NO unit required to generate an N-N bond is energetically extremely unfavourable, which constitutes a general road block for this mechanism. This is also the reason why DNICs are generally stable, and do not show any propensity for N-N bond formation.

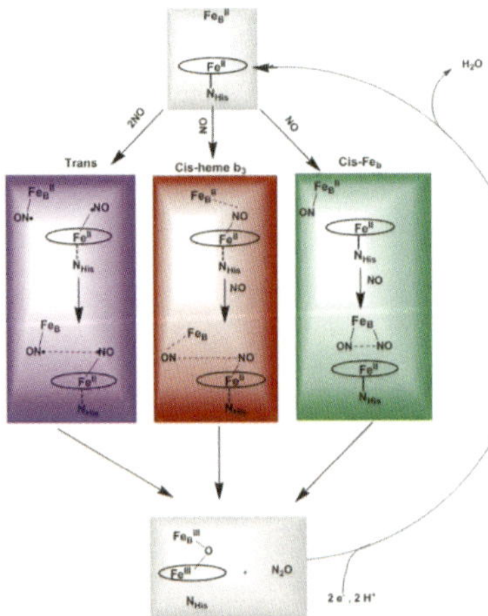

Figure: Three mechanistic possibilities initially proposed for the reduction of NO by NorBC

Fungal NORs (P450nor)

In contrast to the bacterial NORs, which are related to the CCOs, the fungal NORs are derived from Cytochrome P450 and hence, are designated as P450nor. The crystal structure of the enzyme from Fusarium oxysporum shows a heme b with the typical axial cysteine thiolate ligand.

Fungal respiratory NOR: Cyt. P450nor

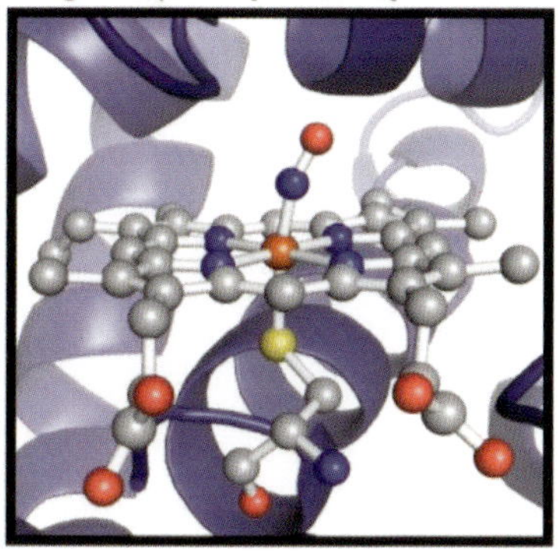

Figure: Ferric NO complex in the active site of P450nor (PDB 1CL6)

Several mechanistic studies have shown that the Fe(III) form is catalytically active in the case of P450nor. Coordination of NO then leads to a six-coordinate low-spin complex that has been characterized with different spectroscopic methods. figure shows a possible mechanism, which is based on kinetic investigations and DFT calculations. In this proposal, the Fe(III)-NO complex reacts with NAD(P)H in an unusual hydride transfer to generate a ferrous heme-NHO complex. Either this species, or the corresponding, protonated hydroxylamide complex corresponds to the reactive “Intermediate I”, which then reacts with a second molecule of NO forming the product N2O. The computational results predict that the imitially formed HNO complex is basic (due to the presence of the axial cysteinate ligand, and that Intermediate I in fact corresponds to the protonated hydroxylamide complex (which can exist in the form of two valence tautomers as indicated in figure).

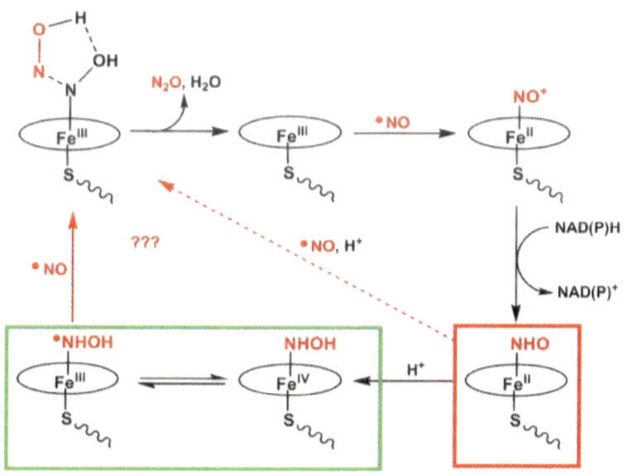

Figure: Proposed mechanism for Cyt. P450nor, based on experimental data combined with DFT calculations.

N2OR

The crystal structure of N2O reductase (N2OR) has been determined for two different organisms. N2OR contains a unique copper cluster, CuZ , which is the designated site of bonding and reduction of the N2O molecule. As shown in figure, the CuZ cluster consists of four copper centres, which are bridged by a sulfide ion. The cluster is connected to the protein through seven His ligands that bind to the Cu centres. Additionally, the CuI-CuIV edge is bridged by a solvent-derived ligand (hydroxide?) in the P. nautica crystal structure, and it has been proposed that this is the site of substrate binding.

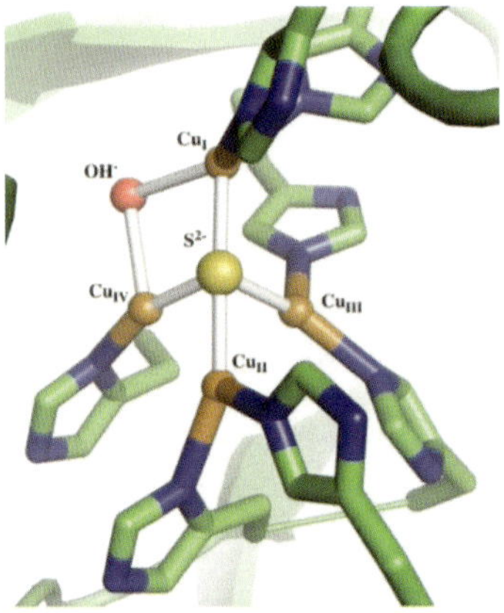

Figure: The active site of N2OR (PDB 1QNI)

Human Impact on Nitrogen Cycle

Impacts on The Atmosphere

One major consequence of human-driven alterations in the nitrogen cycle has been regional and global change in the chemistry of the atmosphere specifically, increased emissions of nitrogen-based trace gases such as nitrous oxide, nitric oxide, and ammonia (NH3). Although such releases have received less attention than increased emissions of carbon dioxide and various sulfur compounds, the trace nitrogen gases cause environmental effects both while airborne and after they are deposited on the ground. For instance, nitrous oxide is long-lived in the atmosphere and contributes to the human-driven enhancement of the greenhouse effect that likely warms the Earth's climate. Nitric oxide is an important precursor of acid rain and photochemical smog. Some of the human activities discussed above affect the atmosphere directly. For instance, essentially all of the more than 20 Tg per year of fixed nitrogen released in automobile exhausts and in other emissions from fossil fuel burning is emitted to the atmosphere as nitric oxide. Other activities indirectly enhance emissions to the atmosphere. Intensive fertilization of agricultural soils can increase the rates at which nitrogen in the form of ammonia is volatilized and lost to the air. It can also speed the microbial breakdown of ammonium and nitrates in the soil, enhancing the release of nitrous oxide. Even in wild or unmanaged lands downwind of agricultural or industrial areas, rain or windborne deposition of human generated nitrogen can spur increased emissions of nitrogen gases from the soils.

Nitrous Oxide

Nitrous oxide is a very effective heat-trapping gas in the atmosphere, in part because it absorbs outgoing radiant heat from the Earth in infrared wavelengths that are not captured by the other major greenhouse gases, water vapour and carbon dioxide. By absorbing and reradiating this heat back toward the Earth, nitrous oxide contributes a few percent to overall greenhouse warming.

Although nitrous oxide is unreactive and long-lived in the lower atmosphere, when it rises into the stratosphere it can trigger reactions that deplete and thin the stratospheric ozone layer that shields the Earth from damaging ultraviolet radiation.

The concentration of nitrous oxide in the atmosphere is currently increasing at the rate of two- to three tenths of a percent per year. While that rise is clearly documented, the sources of the increase remain unresolved. Both fossil fuel burning and the direct impacts of agricultural fertilization have been considered and rejected as the major source. Rather, there is a developing consensus that a wide array of human-driven sources contribute systematically to enrich the terrestrial nitrogen cycle. These dispersed sources include fertilizers, nitrogen-enriched groundwater, nitrogen-saturated forests, forest burning, land clearing, and even the manufacture of nylon, nitric acid, and other industrial products

The net effect is increased global concentrations of a potent greenhouse gas that also contributes to the thinning of the stratospheric ozone layer.

Nitric Oxide and Ammonia

Unlike nitrous oxide, which is unreactive in the lower atmosphere, both nitric oxide and ammonia are highly reactive and therefore much shorter lived. Thus changes in their atmospheric concentrations can be detected only at local or regional scales.

Nitric oxide plays several critical roles in atmospheric chemistry, including catalysing the formation of photochemical (or brown) smog. In the presence of sunlight, nitric oxide and oxygen react with hydrocarbons emitted by automobile exhausts to form ozone, the most dangerous component of smog. Ground-level ozone has serious detrimental effects on human health as well as the health and productivity of crops and forests.

Nitric oxide, along with other oxides of nitrogen and sulfur, can be transformed in the atmosphere into nitric acid and sulfuric acid, which are the major components of acid rain.

Although a number of sources contribute to nitric oxide emissions, combustion is the dominant one. Fossil fuel burning emits more than 20 Tg per year of nitric oxide. Human burning of forests and other plant material may add about 10 Tg, and global emissions of nitric oxide from soils, a substantial fraction of which are human-caused, total 5 to 20 Tg per year. Overall, 80 percent or more of nitric oxide emissions worldwide are generated by human activities, and in many regions the result is increased smog and acid rain.

Human-Caused Global Nitrogen Emissions

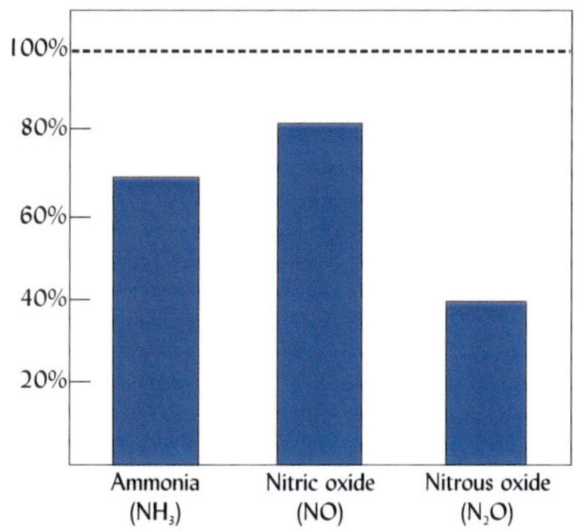

Figure: Human activities are responsible for a large proportion of the global emissions of nitrogen-containing trace gases, including 40% of the nitrous oxide, 80% or more of nitric oxide, and 70% of ammonia releases.

In contrast to nitric oxide, ammonia acts as the primary acid-neutralizing agent in the atmosphere, having an opposite influence on the acidity of aerosols, cloud water, and rainfall. Nearly 70 percent of global ammonia emissions are human-caused. Ammonia volatilized from fertilized fields contributes an estimated 10 Tg per year; ammonia released from domestic animal wastes about 32 Tg; and forest burning some 5 Tg.

The result is increasing atmospheric concentrations of the greenhouse gas nitrous oxide, of the nitrogen precursors of smog, and of biologically available nitrogen that falls from the atmosphere to fertilize ecosystems. Ammonia data are from Schlesinger and Hartley, nitric oxide from Delmas (in press), and nitrous oxide from Prather.

Effects on the Carbon Cycle

Increased emissions of airborne nitrogen have led to enhanced deposition of nitrogen on land and in the oceans. Thanks to the fertilizer effects of nitrogen in stimulating plant growth, this deposition may be acting to influence the atmosphere indirectly by altering the global carbon cycle.

Over much of the Earth's surface, the lushness of plant growth and the accumulation of standing stocks of plant material historically have been limited by scanty nitrogen supplies, particularly in temperate and boreal regions. Human activity has substantially increased the deposition of nitrogen over much of this area, which raises important questions: How much extra plant growth has been caused by human-generated nitrogen additions? As a result, how much extra carbon has been stored in terrestrial ecosystems rather than contributing to the rising concentrations of carbon dioxide in the atmosphere?

Answers to these questions could help explain the imbalance in the carbon cycle that has come to be known as the 'missing sink.' The known emissions of carbon dioxide from human activities such as fossil fuel burning and deforestation exceed by more than 1,000 Tg the amount of carbon dioxide known to be accumulating in the atmosphere each year. Could increased growth rates in terrestrial vegetation be the 'sink' that accounts for the fate of much of that missing carbon?

Experiments in Europe and America indicate that a large portion of the extra nitrogen retained by forest, wetland, and tundra ecosystems stimulates carbon uptake and storage. On the other hand, this nitrogen can also stimulate microbial decomposition and thus releases of carbon from soil organic matter. On balance, however, the carbon uptake through new plant growth appears to exceed the carbon losses, especially in forests.

A number of groups have attempted to calculate the amount of carbon that could be stored in terrestrial vegetation thanks to plant growth spurred by added nitrogen. The resulting estimates range from 100 to 1,300 Tg per year. The number has tended to increase in more recent analyses as the magnitude of human-driven changes in the nitrogen cycle has become clearer. The most recent analysis of the global carbon cycle by the Intergovernmental Panel on Climate Change concluded that nitrogen deposition could represent a major component of the missing carbon sink.

More precise estimates will become possible when we have a more complete understanding of the fraction of human-generated nitrogen that actually is retained within various land-based ecosystems.

Figure: Wild plants living in natural ecosystems, such as this lupine, a nitrogen-fixing plant, dominated the nitrogen cycle for millions of years. Human production of nitrogen fertilizer, burning of fossil fuels, and intensive cultivation of legume crops now adds as much nitrogen to terrestrial ecosystems as do all natural processes combined.

Nitrogen Saturation and Ecosystem Functioning

There are limits to how much plant growth can be increased by nitrogen fertilization. At some point, when the natural nitrogen deficiencies in an ecosystem are fully relieved, plant growth becomes limited by sparsity of other resources such as phosphorus, calcium, or water. When the vegetation can no longer respond to further additions of nitrogen, the ecosystem reaches a state described as nitrogen saturation. In theory, when an ecosystem is fully nitrogen-saturated and its soils, plants, and microbes cannot use or retain any more, all new nitrogen deposits will be dispersed to streams, groundwater, and the atmosphere.

Nitrogen saturation has a number of damaging consequences for the health and functioning of ecosystems. These impacts first became apparent in Europe almost two decades ago when scientists observed significant increases in nitrate concentrations in some lakes and streams and also extensive yellowing and loss of needles in spruce and other conifer forests subjected to heavy nitrogen deposition. These observations led to several field experiments in the U.S. and Europe that have revealed a complex cascade of effects set in motion by excess nitrogen in forest soils.

As ammonium builds up as ammonium builds up in the soil, it is increasingly converted to nitrate by bacterial action, a process that releases hydrogen ions and helps acidify the soil. The build up of nitrate enhances emissions of nitrous oxides from the soil and also encourages leaching of highly water-soluble nitrate into streams or groundwater. As these negatively charged nitrates seep away, they carry with them positively charged alkaline minerals such as calcium, magnesium, and potassium. Thus human modifications to the nitrogen cycle decrease soil fertility by greatly accelerating the loss of calcium and other nutrients that are vital for plant growth. As calcium is depleted and the soil acidified, aluminum ions are mobilized, eventually reaching toxic concentrations that can damage tree roots or kill fish if the aluminum washes into streams. Trees growing in soils replete with nitrogen but starved of calcium, magnesium, and potassium can develop nutrient imbalances in their roots and leaves. This may reduce their photosynthetic rate and efficiency, stunt their growth, and even increase tree deaths.

Nitrogen saturation is much further advanced over extensive areas of northern Europe than in North America because human-generated nitrogen deposition is several times greater there than in even the most extremely affected areas of North America. In the nitrogen-saturated ecosystems of Europe, a substantial fraction of atmospheric nitrate deposits move from the land into streams without ever being taken up by organisms or playing a role in the biological cycle.

In contrast, in the north-eastern U.S., increased leaching of nitrates from the soil and large shifts in the nutrient ratios in tree leaves generally have been observed only in certain types of forests. These include high-elevation sites that receive large nitrogen deposits and sites with shallow soils containing few alkaline minerals to buffer acidification. Elsewhere in the U.S., the early stages of nitrogen saturation have been seen in response to elevated nitrogen deposition in the forests surrounding the Los Angeles Basin and in the Front Range of the Colorado Rockies.

Some forests have a very high capacity to retain added nitrogen, particularly regrowing forests that have been subjected to intense or repeated harvesting, an activity that usually causes severe nitrogen losses. Overall, the ability of a forest to retain nitrogen depends on its potential for further growth and the extent of its current nitrogen stocks. Thus, the impacts of nitrogen deposition are tightly linked to other rapidly changing human-driven variables such as shifts in land use, climate, and atmospheric carbon dioxide and ozone levels.

Effects on Biodiversity and the Species Mix

Limited supplies of biologically available nitrogen are a fact of life in most natural ecosystems, and many native plant species are adapted to function best under this constraint. New supplies of nitrogen showered upon these ecosystems can cause a dramatic shift in the dominant species and also a marked reduction in overall species diversity as the few plant species adapted to take full advantage of high nitrogen out compete their neighbours. In England, for example, nitrogen fertilizers applied to experimental grasslands have led to increased dominance by a few nitrogen-responsive grasses and loss of many other plant species. At the highest fertilization rate, the number of plant species declined more than fivefold. In North America, similarly dramatic reductions in biodiversity have been created by fertilization of grasslands in Minnesota and California. In formerly species-rich heathlands across Western Europe, human-driven nitrogen deposition has been blamed for great losses of biodiversity in recent decades.

In the Netherlands, high human population density, intensive livestock operations, and industries have combined to generate the highest rates of nitrogen deposition in the world. One well-documented consequence has been the conversion of species-rich heathlands to species-poor grasslands and forest. Not only the species richness of the heath but also the biological diversity of the landscape has been reduced because the modified plant communities now resemble the composition of communities occupying more fertile soils. The unique species assemblage adapted to sandy, nitrogen-poor soils is being lost from the region.

Figure: Deposition of nitrogen from the atmosphere is believed to be responsible for the yellowing and loss of needles from conifers and for cases of forest dieback, such as that shown here.

Losses of biodiversity driven by nitrogen deposition can in turn affect other ecological processes. Recent experiments in Minnesota grasslands showed that in ecosystems made species-poor by fertilization, plant productivity was much less stable in the face of a major drought. Even in non-drought years, the normal vagaries of climate produced much more year-to-year variation in the productivity of species-poor grassland plots than in more diverse plots

Figure: Different rates of nitrogen addition lead to marked changes in the plant and insect species compositions and speicies diversity of these plots of grassland vegetation in Minnesota. Each plot is 4m x 4m (about 13 ft x 13 ft), and has received experimental addition of nitrogen (ammonium nitrate) since 1982

Effects on Aquatic Ecosystems

Historical Changes in Water Chemistry

Not surprisingly, nitrogen concentrations in surface waters have increased as human activities have accelerated the rate of fixed nitrogen being put into circulation. A recent study of the North Atlantic Ocean Basin by scientists from a dozen nations estimates that movements of total dissolved nitrogen into most of the temperate-zone rivers in the basin may have increased by two- to

20-fold since preindustrial times (Figure 10). For rivers in the North Sea region, the nitrogen increase may have been six- to 20-fold. The nitrogen increases in these rivers are highly correlated with human generated inputs of nitrogen to their watersheds, and these inputs are dominated by fertilizers and atmospheric deposition.

For decades, nitrate concentrations in many rivers and drinking water supplies have been closely monitored in developed regions of the world, and analysis of these data confirms a historic rise in nitrogen levels in the surface waters. In 1,000 lakes in Norway, for example, nitrate levels doubled in less than a decade. In the Mississippi River, nitrates have more than doubled since 1965. In major rivers of the north-eastern U.S., nitrate concentrations have risen three- to ten-fold since the early 1900s, and the evidence suggests a similar trend in many European rivers.

Again not surprisingly, nitrate concentrations in the world's large rivers rise along with the density of human population in the watersheds. Amounts of total dissolved nitrogen in rivers are also correlated with human population density, but total nitrogen does not increase as rapidly as the nitrate fraction. Evidence indicates that with increasing human disturbance, a higher proportion of the nitrogen in surface waters is composed of nitrate.

Increased concentrations of nitrate have also been observed in groundwater in many agricultural regions, although the magnitude of the trend is difficult to determine in all but a few well-characterized aquifers. Overall, the additions to groundwater probably represent only a small fraction of the increased nitrate transported in surface waters. However, groundwater has a long residence time in many aquifers, meaning that groundwater quality is likely to continue to decline as long as human activities are having substantial impacts on the nitrogen cycle.

High levels of nitrates in drinking water raise significant human health concerns, especially for infants. Microbes in an infant's stomach may convert high levels of nitrate to nitrite. When nitrite is absorbed into the bloodstream, it converts oxygen carrying hemoglobin into an ineffective form called methemoglobin. Elevated methemoglobin levels an anemic condition known as methemoglobinemia can cause brain damage or death. The condition is rare in the U.S., but the potential exists whenever nitrate levels exceed U.S. Public Health Service standards (10 milligrams perliter).

Nitrogen and Acidification of Lakes

Nitric acid is playing an increasing role in the acidification of lakes and streams for two major reasons. One is that most efforts to control acid deposition which includes acid rain, snow, fog, mist, and dry deposits have focused on cutting emissions of sulfur dioxide to limit the formation of sulfuric acid in the atmosphere. In many areas, these efforts have succeeded in reducing inputs of sulfuric acid to soils and water while emissions of nitrogen oxides, the precursors of nitric acid, have gone largely unchecked. The second reason is that many watersheds in areas of moderate to high nitrogen deposition appear to be approaching nitrogen saturation, and the increasingly acidified soils have little capacity to buffer acid rain before it enters streams.

An additional factor in many areas is that nitric acid predominates among the pollutants that accumulate in the winter snowpack. Much of this nitric acid is flushed out with the first batch of spring melt water, often washing a sudden, concentrated acid pulse into vulnerable lakes.

Adding inorganic nitrogen to freshwater ecosystems that are also rich in phosphorus can eutrophy as well as acidify the waters. Both eutrophication and acidification generally lead to decreased diversity of both plant and animal species. Fish populations, in particular, have been reduced or eliminated in many acidified lakes across Scandinavia, Canada, and north-eastern United States.

Figure: Native grasslands in Minnesota often contain 20 to 30 or more plant species per square meter, as does this plot. This plot is a control plot that received no nitrogen, and that retained its original plant diversity.

Because the extent of nitrogen-saturated ecosystems continues to grow, along with human-caused nitrogen deposition, controls on sulfur dioxide emissions alone clearly will not be sufficient to decrease acid rain or prevent its detrimental effects on streams and lakes. European governments already have recognized the importance of nitrogen in acidifying soils and waters, and intergovernmental efforts are underway there to reduce emissions and deposition of nitrogen on a regional basis.

Eutrophication in Estuaries and Coastal Waters

One of the best documented and best understood consequences of human alterations of the nitrogen cycle is the eutrophication of estuaries and coastal seas. It is arguably the most serious human threat to the integrity of coastal ecosystems.

In sharp contrast to the majority of temperate zone lakes, where phosphorus is the nutrient that most limits primary productivity by algae and other aquatic plants and controls eutrophication, these processes are controlled by nitrogen inputs in most temperate-zone estuaries and coastal waters. This is largely because the natural flow of nitrogen into these waters and the rate of nitrogen fixation by planktonic organisms are relatively low while microbes in the sea floor sediments actively release nitrogen back to the atmosphere.

When high nitrogen loading causes eutrophication in stratified waters where a sharp temperature gradient prevents mixing of warm surface waters with colder bottom waters the result can be anoxia (no oxygen) or hypoxia (low oxygen) in bottom waters. Both conditions appear to be becoming more prevalent in many estuaries and coastal seas. There is good evidence that since the 1950s or 1960s, anoxia has increased in the Baltic Sea, the Black Sea, and Chesapeake Bay. Periods of hypoxia have increased in Long Island Sound, the North Sea, and the Kattegat, resulting in significant losses of fish and shellfish.

Figure: Nitrogen addition to this plot, located near that shown in Figure, led to the loss of almost all native prairie species and to dominance by the weedy European quack grass. In 1982 this plot looked much like the one shown in Figure.

Eutrophication is also linked to losses of diversity, both in the sea floor community including seaweeds, sea grasses, and corals and among planktonic organisms. In eutrophied waters, for example, nuisance algae may come to dominate the phytoplankton community. Increases in troublesome or toxic algal blooms have been observed in many estuaries and coastal seas worldwide in recent decades. During the 1980s, toxic blooms of dinoflagellates and brown-tide organisms caused extensive die-offs of fish and shellfish in many estuaries. Although the causes are not completely understood, there is compelling evidence that nutrient enrichment of coastal waters is at least partly to blame for such blooms.

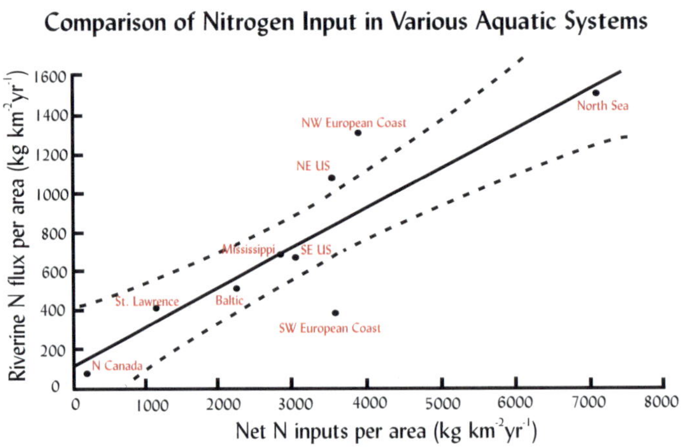

Figure: Movements of nitrogen into most of the temperate zone rivers that empty into the North Atlantic Ocean have increased by two to 20-fold since preindustrial times.

Nitrogen increases in these rivers are highly correlated with increasing human generated nitrogen inputs into the watersheds, particularly fertilizer use and rising atmospheric deposition of nitrogen. Modified from Howarth.

Major Uncertainties

Although this report has focused on what is known about human-driven changes to the global nitrogen cycle, major uncertainties remain. Some of these have been noted in earlier sections. This topic, however, focuses on important processes that remain so poorly understood that it is difficult to distinguish human-caused impacts or to predict their consequences.

Marine Nitrogen Fixation

Little is known about the unmodified nitrogen cycle in the open ocean. Credible estimates of nitrogen fixation by organisms in the sea vary more than ten-fold, ranging from less than 30 to more than 300 Tg per year. There is some evidence that human alteration of the nitrogen cycle could alter biological processes in the open ocean, but there is no adequate frame of reference against which to evaluate any potential human-driven change in marine nitrogen fixation.

Changes in Limiting Resources

One consequence of human-driven changes in the global nitrogen cycle is a shift in the resources that limit biological processes in many areas. Large amounts of nitrogen are now deposited on many ecosystems that were once nitrogen deficient. The dominant species in these systems may have evolved with nitrogen limitation, and the ways they grow and function and form symbiotic partnerships may reflect adaptations to this limit. With this limit removed, species must operate under novel constraints such as now-inadequate phosphorus or water supplies. How are the performance of organisms and the operation of larger ecological processes affected by changes in their chemical environment for which they have no evolutionary background and to which they are not adapted?

Capacity to Retain Nitrogen

Forests and wetlands vary substantially in their capacity to retain added nitrogen. Interacting factors that are known to affect this capacity include soil texture, degree of chemical weathering of soil, fire history, rate at which plant material accumulates, and past human land use. However, we still lack a fundamental understanding of how and why nitrogen-retention processes vary among ecosystems much less how they have changed and will change with time.

Figure: The bottom-dwelling plants of a marine ecosystem that received natural rates of nitrogen addition. Note the high diversity of these plants and their spacing.

Alteration of Denitrification

In large river basins, the majority of nitrogen that arrives is probably broken down by denitrifying bacteria and released to the atmosphere as nitrogen gas or nitrous oxide. Exactly where most of this activity takes place is poorly understood, although we know that streamside areas and wetlands are important. Human activities such as increased nitrate deposition, dam building, and rice cultivation have probably enhanced denitrification, while draining of wetlands and alteration of riparian ecosystems has probably decreased it. But the net effect of human influence remains uncertain.

Natural Nitrogen Cycling

Information on the rate of nitrogen deposition and loss in various regions prior to extensive human alterations of the nitrogen cycle remains patchy. In part, this reflects the fact that all of the Earth already is affected to some degree by human activity. Nevertheless, studies in remote regions of the Southern Hemisphere illustrate that there is still valuable information to be gathered on areas that have been minimally altered by humans.

Future Prospects and Management Options

Fertilizer use

The greatest human-driven increases in global nitrogen supplies are linked to activities intended to boost food production. Modern intensive agriculture requires large quantities of nitrogen fertilizer; humanity, in turn, requires intensive agriculture to support a growing population that is projected to double by the end of the next century. Consequently, the production and application of nitrogen fertilizer has grown exponentially, and the highest rates of application are now found in some developing countries with the highest rates of population growth. One study predicts that by the year 2020, global production of nitrogen fertilizer will increase from a current level of about 80 Tg to 134 Tg per year.

Curtailing this growth in nitrogen fertilizer production will be a difficult challenge. Nevertheless, there are ways to slow the growth of fertilizer use and also to reduce the mobility and hence the regional and global impacts of the nitrogen that is applied to fields.

One way to reduce the amount of fertilizer used is to increase its efficiency. Often at least half of the fertilizer applied to fields is lost to the air or water. This leakage represents an expensive waste to the farmer as well as a significant driver of environmental change. A number of management practices have been identified that can reduce the amounts of fertilizer used and cut losses of nitrogen to the air and water without sacrificing yields or profits (and in some cases, increasing them). For instance, one commercial sugar cane plantation in Hawaii was able to cut nitrogen fertilizer use by one third and reduce losses of nitrous oxide and nitric oxide tenfold by dissolving the fertilizer in irrigation water, delivering it below the soil surface, and timing multiple applications to meet the needs of the growing crop. This knowledge-intensive system also proved more profitable than broadcasting fewer, larger applications of fertilizer onto the soil surface. The widespread implementation of such practices, particularly in developing regions, should be a high priority for agronomists as well as ecologists since improved practices provide an opportunity to reduce the costs of food production while slowing the rate of global change.

There are also ways to prevent the nitrogen that leaches from fertilized farmland from reaching streams, estuaries and coastal waters where it contributes to eutrophication. In many regions, agricultural lands have been expanded by channelizing streams, clearing riparian forests, and draining wetlands. Yet these areas serve as important natural nitrogen traps. Restoration of wetlands and riparian areas and even construction of artificial wetlands have been shown to be effective in preventing excess nitrogen from entering waters.

Figure: The bottom-dwelling plants of a marine ecosystem that received high rates of nitrogen input. Note that there are few plant species, and that the leaves of these are covered with a thick layer of algae.

Figure: Human activities, such as fertilizer production, growing legume crops, and burning of fossil fuels, are now of equal or greater magnitude than natural processes in the nitrogen cycle.

Human domination of the nitrogen cycle impacts the functioning of many terrestrial and aquatic ecosystems, including seemingly pristine habitats such as this alpine ecosystem.

Fossil Fuel Burning

The second major source of human-fixed nitrogen is fossil fuel burning. It, too, will increase markedly as we enter the next century, particularly in the developing world. One study predicts that production of nitrogen oxides from fossil fuels will more than double in the next 25 years, from about 20 Tg per year to 46 Tg. Reducing these emissions will require improvements in the efficiency of fuel combustion as well as in the interception of airborne by products of combustion. As with improvements in fertilizer efficiency, it will be particularly important to transfer efficient combustion technologies to developing countries as their economies and industries grow.

Rock Cycle

Rocks are the most common material on Earth. They are naturally occurring aggregates of one or more minerals.

Rock divisions occur in three major families based on how they formed: igneous, sedimentary, and metamorphic. Each group contains a collection of rock types that differ from each other on the basis of the size, shape, and arrangement of mineral grains.

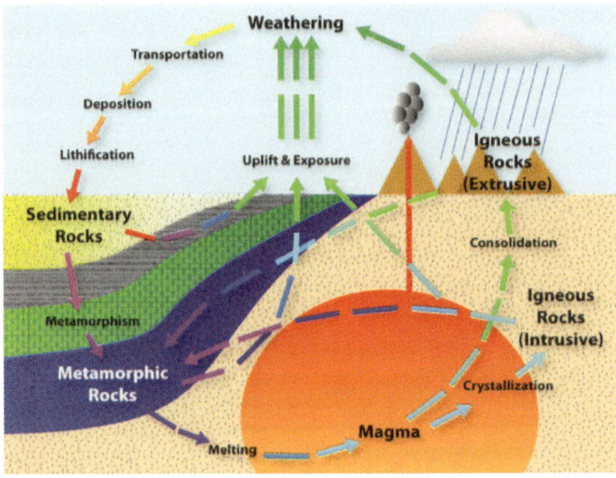

The rock cycle is an illustration that is used to explain how the three rock types are related to each other and how Earth processes change a rock from one type to another through geologic time. Plate tectonic movement is responsible for the recycling of rock materials and is the driving force of the rock cycle.

Processes of Rock Cycle

Transition to Igneous Rock

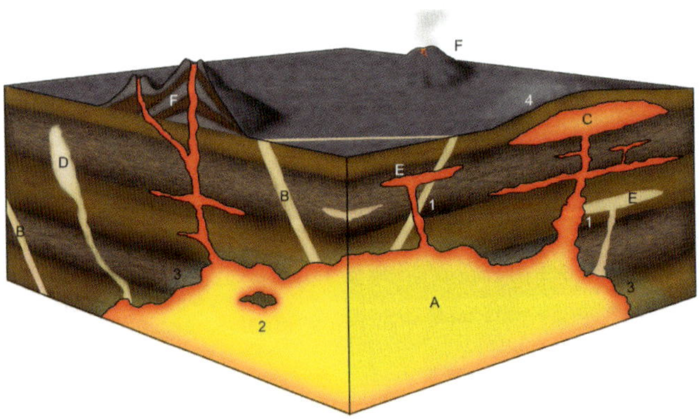

Structures of Igneous Rock. Legend: A = magma chamber (batholith); B = dyke/dike; C = laccolith; D = pegmatite; E = sill; F = stratovolcano; processes: 1 = newer intrusion cutting through older one; 2 = xenolith or roof pendant; 3 = contact metamorphism; 4 = uplift due to laccolith emplacement.

When rocks are pushed deep under the Earth's surface, they may melt into magma. If the conditions no longer exist for the magma to stay in its liquid state, it cools and solidifies into an igneous rock. A rock that cools within the Earth is called intrusive or plutonic and cools very slowly, producing a coarse-grained texture such as the rock granite. As a result of volcanic activity, magma (which is called lava when it reaches Earth's surface) may cool very rapidly while being on the Earth's surface exposed to the atmosphere and are called extrusive or volcanic rocks. These rocks are fine-grained and sometimes cool so rapidly that no crystals can form and result in a natural glass, such as obsidian, however the most common fine grained rock would be known as basalt. Any of the three main types of rocks (igneous, sedimentary, and metamorphic rocks) can melt into magma and cool into igneous rocks.

Secondary Changes

Epigenetic change (secondary processes occurring at low temperatures and low pressures) may be arranged under a number of headings, each of which is typical of a group of rocks or rock-forming minerals, though usually more than one of these alterations is in progress in the same rock. Silicification, the replacement of the minerals by crystalline or crypto-crystalline silica, is most common in felsic rocks, such as rhyolite, but is also found in serpentine, etc. Kaolinization is the decomposition of the feldspars, which are the most common minerals in igneous rocks, into kaolin (along with quartz and other clay minerals); it is best shown by granites and syenites. Serpentinization is the alteration of olivine to serpentine (with magnetite); it is typical of peridotites, but occurs in most of the mafic rocks. In uralitization, secondary hornblende replaces augite; chloritization is the alteration of augite (biotite or hornblende) to chlorite, and is seen in many diabases, diorites and greenstones. Epidotization occurs also in rocks of this group, and consists in the development of epidote from biotite, hornblende, augite or plagioclase feldspar.

This diamond is a mineral from within an igneous or metamorphic rock that formed at high temperature and pressure.

Transition to Metamorphic Rock

Rocks exposed to high temperatures and pressures can be changed physically or chemically to form a different rock, called metamorphic. Regional metamorphism refers to the effects on large masses of rocks over a wide area, typically associated with mountain building events within orogenic belts. These rocks commonly exhibit distinct bands of differing mineralogy and colors, called foliation. Another main type of metamorphism is caused when a body of rock comes into contact with an igneous intrusion that heats up this surrounding country rock. This *contact metamorphism* results in a rock that is altered and re-crystallized by the extreme heat of the magma and/or by the addition of fluids from the magma that add chemicals to the surrounding rock (metasomatism). Any pre-existing type of rock can be modified by the processes of metamorphism.

Transition to Sedimentary Rock

Rocks exposed to the atmosphere are variably unstable and subject to the processes of weathering and erosion. Weathering and erosion break the original rock down into smaller fragments and carry away dissolved material. This fragmented material accumulates and is buried by additional material. While an individual grain of sand is still a member of the class of rock it was formed from, a rock made up of such grains fused together is sedimentary. Sedimentary rocks can be formed from the lithification of these buried smaller fragments (clastic sedimentary rock), the accumulation and lithification of material generated by living organisms (biogenic sedimentary rock - fossils), or lithification of chemically precipitated material from a mineral bearing solution due to evaporation (precipitate sedimentary rock). Clastic rocks can be formed from fragments broken apart from larger rocks of any type, due to processes such as erosion or from organic material, like plant remains. Biogenic and precipitate rocks form from the deposition of minerals from chemicals dissolved from all other rock types.

Forces that Drive the Rock Cycle

Plate Tectonics

In 1967, J. Tuzo Wilson describing the repeated opening and closing of ocean basins, in particular focusing on the current Atlantic Ocean area. This concept, a part of the plate tectonics revolution, became known as the *Wilson cycle*. The Wilson cycle has had profound effects on the modern interpretation of the rock cycle as plate tectonics became recognized as the driving force for the rock cycle.

Spreading Ridges

At the mid-ocean divergent boundaries new magma is produced by mantle upwelling and a shallow melting zone. This juvenile basaltic magma is an early phase of the igneous portion of the cycle. As the tectonic plates on either side of the ridge move apart the new rock is carried away from the ridge, the interaction of heated circulating seawater through fractures starts the retrograde metamorphism of the new rock.

Subduction Zones

The new basaltic oceanic crust eventually meets a subduction zone as it moves away from the spreading ridge. As this crust is pulled back into the mantle, the increasing pressure and temperature conditions cause a restructuring of the mineralogy of the rock, this metamorphism alters the rock to form eclogite. As the slab of basaltic crust and some included sediments are dragged deeper, water and other more volatile materials are driven off and rise into the overlying wedge of rock above the subduction zone, which is at a lower pressure. The lower pressure, high temperature, and now volatile rich material in this wedge melts and the resulting buoyant magma rises through the overlying rock to produce island arc or continental margin volcanism. This volcanism includes more silicic lavas the further from the edge of the island arc or continental margin, indicating a deeper source and a more differentiated magma.

At times some of the metamorphosed downgoing slab may be thrust up or obducted onto the continental margin. These blocks of mantle peridotite and the metamorphic eclogites are exposed as ophiolite complexes.

The newly erupted volcanic material is subject to rapid erosion depending on the climate conditions. These sediments accumulate within the basins on either side of an island arc. As the sediments become more deeply buried lithification begins and sedimentary rock results.

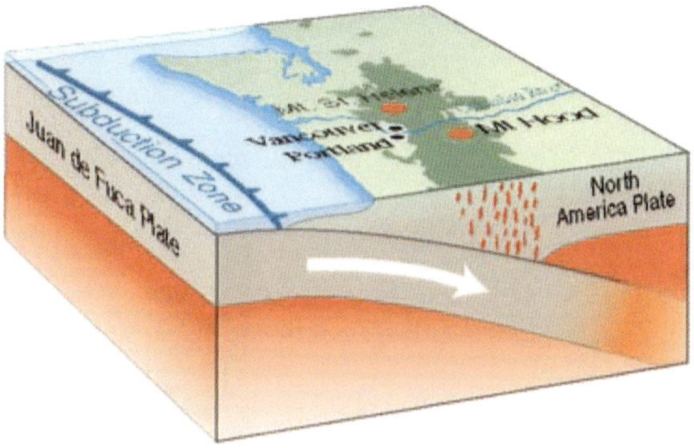

The Juan de Fuca plate sinks below the North America plate at the Cascadia subduction zone.

Continental Collision

On the closing phase of the classic Wilson cycle, two continental or smaller terranes meet at a convergent zone. As the two masses of continental crust meet, neither can be subducted as they are both *low density* silicic rock. As the two masses meet, tremendous compressional forces distort and modify the rocks involved. The result is regional metamorphism within the interior of the ensuing orogeny or mountain building event. As the two masses are compressed, folded and faulted into a mountain range by the continental collision the whole suite of pre-existing igneous, volcanic, sedimentary and earlier metamorphic rock units are subjected to this new metamorphic event.

Accelerated Erosion

The high mountain ranges produced by continental collisions are immediately subjected to the forces of erosion. Erosion wears down the mountains and massive piles of sediment are developed in adjacent ocean margins, shallow seas, and as continental deposits. As these sediment piles are buried deeper they become lithified into sedimentary rock. The metamorphic, igneous, and sedimentary rocks of the mountains become the new piles of sediments in the adjoining basins and eventually become sedimentary rock.

An Evolving Process

The plate tectonics rock cycle is an evolutionary process. Magma generation, both in the spreading ridge environment and within the wedge above a subduction zone, favours the eruption of the more silicic and volatile rich fraction of the crustal or upper mantle material. This lower density material tends to stay within the crust and not be subducted back into the mantle. The magmatic aspect of plate tectonics tends to gradual segregation within or between the mantle and crust. As magma forms, the initial melt is composed of the more silicic phases that have a lower melting point. This leads to partial melting and further segregation of the lithosphere. In addition the silicic continental crust is relatively buoyant and is not normally subducted back into the mantle. So over time the continental masses grow larger and larger.

The Role of Water

The presence of abundant water on Earth is of great importance for the rock cycle. Most obvious perhaps are the water driven processes of weathering and erosion. Water in the form of precipitation and acidic soil water and groundwater is quite effective at dissolving minerals and rocks, especially those igneous and metamorphic rocks and marine sedimentary rocks that are unstable under near surface and atmospheric conditions. The water carries away the ions dissolved in solution and the broken down fragments that are the products of weathering. Running water carries vast amounts of sediment in rivers back to the ocean and inland basins. The accumulated and buried sediments are converted back into rock.

A less obvious role of water is in the metamorphism processes that occur in fresh seafloor volcanic rocks as seawater, sometimes heated, flows through the fractures and crevices in the rock. All of these processes, illustrated by serpentinization, are an important part of the destruction of volcanic rock.

The role of water and other volatiles in the melting of existing crustal rock in the wedge above a subduction zone is a most important part of the cycle. Along with water, the presence of carbon dioxide and other carbon compounds from abundant marine limestone within the sediments atop the down going slab is another source of melt inducing volatiles. This involves the carbon cycle as a part of the overall rock cycle.

Climate Influences on Rock Cycle

Climate has three basic influences on the rock cycle: weathering & erosion, rate of limestone production, and rate of fossil production.

Weathering and erosion: Weathering is the breakup of rock due to physical and chemical processes, while erosion is the transport of weathered rock particles by wind and water. The primary way that climate affects weathering and erosion through its connection to the water cycle.

When water freezes in cracks and crevices in rock and expands, the rocks are physically broken apart. The amount of rainfall will affect weathering as will the types of plants in the area (which are greatly affected by the amount of rainfall), whose roots break rocks apart and may chemically dissolve some rocks.

Rate of limestone production: Limestone rocks are made most prolifically in the warm shallow waters of tropics. Thus, limestone production depends on the extent of tropical regions. The amount of shallow sea area, where limestone forms quickly, depends on sea level. When the edges of the continents are flooded by high sea level because glaciers have melted during warm global climate, more carbonate rocks are produced.

In warm, tropical oceans, like that shown in (A), large numbers of corals and other marine animals and plants make skeletons out of calcite and other carbonate minerals.

These skeletons and carbonate mud make a rock called limestone like the one shown in (B) from San Salvador Island in the Bahamas. This limestone was a coral reef living under a shallow sea about 120,000 years ago.

The leg and shoulder bones of a large Sauropod dinosaur preserved at Dinosaur National Monument (Colorado, USA) Credits: Courtesy of Dinosaur National Monument.

Rate of fossil production: Environments and ecosystems depend on climate so different types of sedimentary rocks and fossils are preserved when climates change. Additionally, fossilization processes depend on climate. The chance of animal or plant remains becoming fossilized at all is very minimal, and it is much more likely that the remains will decompose.

Body fossils are remains of actual organisms. Most living things never become fossils. It takes special conditions for a fossil to form. Hard parts made of mineral such as shells and bones are much more likely to become body fossils than soft tissues, such as skin, organs, and eyes, which usually decay. This means that animals like jellyfish, which have no bones made of hard mineral, are rarely preserved.

But since the rate of decomposition is different in different climates, remains are more likely to become fossils in some environment than others. In warm, humid environments, less soft tissues are fossilized because the rate of decomposition is high. However, in arctic environments, much like in your freezer, the rate of decomposition is lower and soft tissues are more likely to fossilize.

Energy Cycle

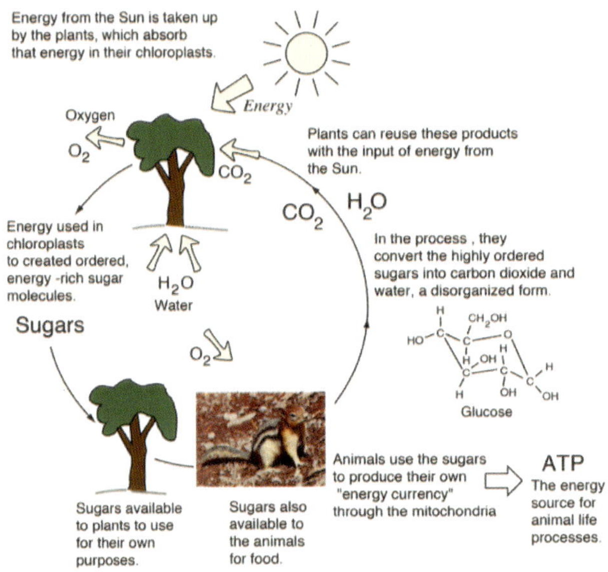

A fascinating parallel between plant and animal life is in the use of tiny energy factories within the cells to handle the energy transformation processes necessary for life. In plants, these energy factories are called chloroplasts. They collect energy from the sun and use carbon dioxide and water in the process called photosynthesis to produce sugars. Animals can make use of the sugars provided by the plants in their own cellular energy factories, the mitochondria. These produce a versatile energy currency in the form of adenosine triphosphate (ATP). This high-energy molecule stores the energy we need to do just about everything we do.

The energy cycle for life is fueled by the Sun. The main end product for plants and animals is the production of highly energetic molecules like ATP. These molecules store enough immediately available energy to allow plants and animals to do their necessary work.

The Work of Living Things

There are a number of energy transformations in plants and animals which are essential to life. These processes will be described as the "work" of living things, although it is not immediately evident that they involve work in the physical sense. Some of the general kinds of processes involved in the energy cycle are:

Synthetic Work

Both plants and animals must make the complex molecules necessary for life. One example is the production of DNA - your genetic material. If you don't make extra copies of DNA, you will have no information to pass on to your offspring. Every time one of your cells makes a copy of itself, it has to make a new copy of the DNA. That synthesis process requires a lot of ATP.

The process of growth requires a lot of synthetic work to create the new cells and enlarge the structures.

Electrical Work

You may not think of yourself as an electrically operated machine, but you are. Each of our cells has an electric potential associated with it. This potential, or voltage, helps to control the migration of ions across the cell membranes. A major example of electrical work is in the operation of the nerves. When your nerves fire, they generate an electrical impulse called an action potential which can communicate information to your brain, or carry a signal from your brain to a muscle to initiate its movement.

Electrical energy transformation is essential for sensing your environment as well as for reacting to that environment in any way.

Mechanical Work

Most easily visualized is the mechanical work associated with the moving of our muscles. This muscle movement is very important and requires a lot of energy. The source of that energy is ATP.

References

- Blatt, Harvey & Robert J. Tracy (1996). Petrology; Igneous, Sedimentary, and Metamorphic, 2nd Ed. W. H. Freeman. ISBN 0-7167-2438-3
- Water-cycle, science: britannica.com, Retrieved 28 April 2018
- What-human-activities-affect-the-carbon-cycle-12083853: sciencing.com, Retrieved 18 July 2018
- Nitrogen-assimilation-in-plants-201012251621: preservearticles.com, Retrieved 11 March 2018
- Fichter, Lynn S., (2000), The Wilson Cycle and a Plate Tectonic Rock Cycle, James Madison University, Department of Geology and Environmental Science. Retrieved 18 Aug. 2005
- Water-cycle: conserve-energy-future.com, Retrieved 11 June 2018
- Nitrogen-cycle, science: britannica.com, Retrieved 28 May 2018

4

Ecosystem and its Types

A community which consists of living organisms and non living components like air, soil and water is referred to as an ecosystem. It can be divided into various types such as aquatic ecosystem, terrestrial ecosystem, etc. The varied categories of ecosystems and their related aspects have been carefully analyzed in this chapter.

Ecosystem

An ecosystem, a term very often used in biology, is a community of plants and animals interacting with each other in a given area, and also with their non-living environments. The non-living environments include weather, earth, sun, soil, climate and atmosphere. The ecosystem relates to the way that all these different organisms live in close proximity to each other and how they interact with each other. For instance, in an ecosystem where there are both rabbits and foxes, these two creatures are in a relationship where the fox eats the rabbit in order to survive. This relationship has a knock on effect with the other creatures and plants that live in the same or similar areas. For instance, the more rabbits that foxes eat, the more the plants may start to thrive because there are fewer rabbits to eat them.

Ecosystems can be huge, with many hundreds of different animals and plants all living in a delicate balance, or they could be relatively small. In particularly harsh places in the world, particularly the North and South Poles, the ecosystems are relatively simple because there are only a few types of creatures that can withstand the freezing temperatures and harsh living conditions. Some creatures can be found in multiple different ecosystems all over the world in different relationships with other or similar creatures. Ecosystems also consist of creatures that mutually benefit from each other. For instance, a popular example is that of the clown fish and the anemone – the clown fish cleans the anemone and keeps it safe from parasites as the anemone stings bigger predators that would otherwise eat clown fish.

> Earth as an ecosystem stands out in the all of the universe. There's no place that we know about that can support life as we know it, not even our sister planet, mars, where we might set up housekeeping someday, but at great effort and trouble we have to recreate the things we take for granted here.
>
> – Sylvia earle

An ecosystem can be destroyed by a stranger. The stranger could be rise in temperature or rise in sea level or climate change. The stranger can affect the natural balance and can harm or destroy the ecosystem. It's a bit unfortunate but ecosystems have been destroyed and vanished by man-made activities like deforestation, urbanization and natural activities like floods, storms, fires or volcanic eruptions.

Ecosystem Structure

At a basic functional level, ecosystem generally contains primary producers (plants) capable of harvesting energy from the sun through the process called photosynthesis. This energy then flows through the food chain. Next come consumers. Consumers could be primary consumers (herbivores) or secondary consumers (carnivores). These consumers feed on the captured energy. Decomposers work at the bottom of the food chain. Dead tissues and waste products are produced at all levels. Scavengers, detritivores and decomposers not only feed on this energy but also break organic matter back into its organic constituents. It is the microbes that finish the job of decomposition and produce organic constituents that can again be used by producers.

Energy that flows through the food chain i.e. from producers to consumers to decomposers is always inefficient. That means less energy is available at secondary consumers level than at primary producers level. It's not surprising but amount of energy produced from place to place varies a lot due to amount of solar radiation and the availability of nutrients and water.

Types of Ecosystem

There are very many types of ecosystems out there, but the three major classes of ecosystems, sometimes referred to as 'biomes', which are relatively contained, are the following:

- Freshwater Ecosystems
- Terrestrial Ecosystems
- Ocean Ecosystems

Freshwater Ecosystems

These can then be broken up into smaller ecosystems. For instance, in the freshwater ecosystems we find:

- Pond Ecosystems: These are usually relatively small and contained. Most of the time they include various types of plants, amphibians and insects. Sometimes they include fish, but as these cannot move around as easily as amphibians and insects, it is less likely, and most of the time fish are artificially introduced to these environments by humans.
- River Ecosystems: Because rivers always link to the sea, they are more likely to contain fish alongside the usual plants, amphibians and insects.

These sorts of ecosystems can also include birds because birds often hunt in and around water for small fish or insects.

As is clear from the title, freshwater ecosystems are those that are contained to freshwater environments. This includes, but is not limited to, ponds, rivers and other waterways that are not the sea (which is, of course, saltwater and cannot support freshwater creatures for very long). Freshwater ecosystems are actually the smallest of the three major classes of ecosystems, accounting for just 1.8% of the total of the Earth's surface. The ecosystems of freshwater systems include relatively small fish (bigger fish are usually found in the sea), amphibians (such as frogs, toads and newts), insects of various sorts and, of course, plants. The absolutely smallest living part of the food web of these sorts of ecosystems is plankton, a small organism that is often eaten by fish and other small creatures.

Terrestrial Ecosystems

Terrestrial ecosystems are many because there are so many different sorts of places on Earth. Some of the most common terrestrial ecosystems that are found are the following:

- Rainforests – Rainforests usually have extremely dense ecosystems because there are so many different types of animals all living in a very small area.
- Tundra – As mentioned above, tundra usually have relatively simple ecosystems because of the limited amount of life that can be supported in these harsh conditions.
- Deserts – Quite the opposite of tundra in many ways, but still harsh, more animals live in the extreme heat than live in the extreme cold of Antarctica, for instance.
- Savannas – These differ from deserts because of the amount of rain that they get each year. Whereas deserts get only a tiny amount of precipitation every tea, savannas tend to be a bit wetter which is better for supporting more life.
- Forests – There are many different types of forests all over the world including deciduous forests and coniferous forests. These can support a lot of life and can have very complex ecosystems.
- Grasslands – Grasslands support a wide variety of life and can have very complex and involved ecosystems.

Since there are so many different types of terrestrial ecosystems, it can be difficult to make generalizations that cover them all.

Because terrestrial ecosystems are so diverse, it is difficult to make generalizations about them. However, a few things are true almost all of the time. For instance, most contain herbivores that eat plants (that get their sustenance from the sun and the soil) and all have carnivores that eat herbivores and other carnivores. Some places, such the poles, contain mainly carnivores because not plant life grows. A lot of animals and plants that grow and live in terrestrial ecosystems also interact with freshwater and sometimes even ocean ecosystems.

Ocean Ecosystems

Ocean ecosystems are relatively contained, although they, like freshwater ecosystems, also include certain birds that hunt for fish and insects close to the ocean's surface. There are different sorts of ocean ecosystems:

- Shallow water: Some tiny fish and coral only live in the shallow waters close to land.
- Deep water: Big and even gigantic creatures can live deep in the waters of the oceans. Some of the strangest creatures in the world live right at the bottom of the sea.
- Warm water: Warmer waters, such as those of the Pacific Ocean, contain some of the most impressive and intricate ecosystems in the world.

- Cold water: Less diverse, cold waters still support relatively complex ecosystems. Plankton usually form the base of the food chain, following by small fish that are either eaten by bigger fish or by other creatures such as seals or penguins.

Ocean ecosystems are amongst some of the most interesting in the world, especially in warm waters such as those of the Pacific Ocean. This is not least because around 75% of the Earth is covered by the sea, which means that there is lots of space for all sorts of different creatures to live and thrive. There are actually three different types of oceanic ecosystems: shallow waters, deep waters and the deep ocean surface. In two of these the very base of the food chain is plankton, just as it is in freshwater ecosystems.

These plankton and other plants that grow in the ocean close to the surface are responsible for 40% of all photosynthesis that occurs on Earth. From this there are herbivorous creatures that eat the plankton, such as shrimp, that are then themselves usually eaten by bigger creatures, particularly fish. Interesting, in the deep ocean, plankton cannot exist because photosynthesis cannot occur since light cannot penetrate that far into the ocean's depths. Down in the deepest depths of the ocean, therefore, creatures have adapted very strangely and are amongst some of the most fascination and the most terrifying and intriguing living creatures on Earth.

Energy Flow

Energy has been defined as the capacity to do work. Energy exists in two forms potential and kinetic.

Potential energy is the energy at rest (i.e., stored energy) capable of performing work. Kinetic energy is the energy of motion (free energy).

It results in work performance at the expense of potential energy. Conversion of potential energy into kinetic energy involves the imparting of motion.

The source of energy required by all living organisms is the chemical energy of their food. The chemical energy is obtained by the conversion of the radiant energy of sun.

The radiant energy is in the form of electromagnetic waves which are released from the sun during the transmutation of hydrogen to helium. The chemical energy stored in the food of living organisms is converted into potential energy by the arrangement of the constituent atoms of food in a particular manner. In any ecosystem there should be unidirectional flow of energy.

This energy flow is based on two important Laws of Thermodynamics which are as follows:

(1) The first law of Thermodynamics

It states that the amount of energy in the universe is constant. It may change from one form to another, but it can neither be created nor destroyed. Light energy can be neither created nor destroyed as it passes through the atmosphere. It may, however, be transformed into another type of energy, such as chemical energy or heat energy. These forms of energy cannot be transformed into electromagnetic radiation.

(2) The second law of Thermodynamics

It states that non-random energy (mechanical, chemical, radiant energy) cannot be changed without some degradation into heat energy. The change of energy from one form to another takes place in such a way that a part of energy assumes waste form (heat energy). In this way, after transformation the capacity of energy to perform work is decreased. Thus, energy flows from higher to lower level.

Main source of energy is sun. Approximately 57% of sun energy is absorbed in the atmosphere and scattered in the space. Some 35% is spent to heat water and land areas and to evaporate water. Of the approximately 8% of light energy striking plant surface, 10% to 15% is reflected, 5% is transmitted and 80 to 85% is absorbed; and an average of only 2% (0.5 to 3.5%) of the total light energy striking on a leaf is used in photosynthesis and rest is transformed into heat energy.

Energy Flow in Ecosystems

Living organisms can use energy in two forms radiant and fixed energy. Radiant energy is in the form of electromagnetic waves, such as light. Fixed energy is potential chemical energy bound in various organic substances which can be broken down in order to release their energy content.

Organisms that can fix radiant energy utilizing inorganic substances to produce organic molecules are called autotrophs. Organisms that cannot obtain energy from abiotic source but depend on energy-rich organic molecules synthesized by autotrophs are called heterotrophs. Those which obtain energy from living organisms are called consumers and those which obtain energy from dead organisms are called decomposers.

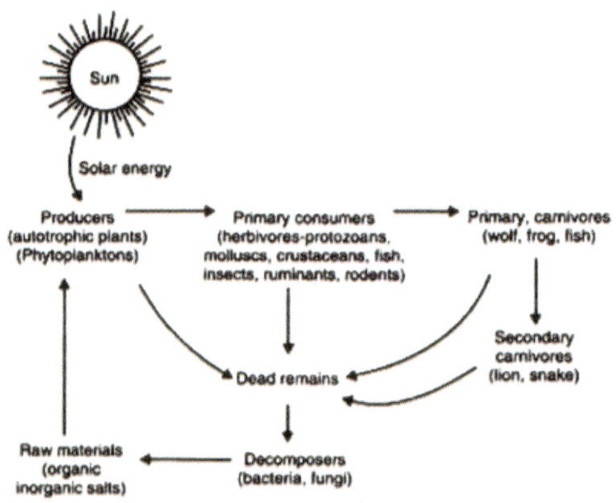

Flow of energy at different levels of ecosystem.

When the light energy falls on the green surfaces of plants, a part of it is transformed into chemical energy which is stored in various organic products in the plants. When the herbivores consume plants as food and convert chemical energy accumulated in plant products into kinetic energy, degradation of energy will occur through its conversion into heat. When herbivores are consumed by carnivores of the first order (secondary consumers) further degradation will occur. Similarly, when primary carnivores are consumed by top carnivores, again energy will be degraded.

Trophic Level

The producers and consumers in ecosystem can be arranged into several feeding groups, each known as trophic level (feeding level). In any ecosystem, producers represent the first trophic level, herbivores present the second trophic level, primary carnivores represent the third trophic level and top carnivores represent the last level.

Food Chain

In the ecosystem, green plants alone are able to trap in solar energy and convert it into chemical energy. The chemical energy is locked up in the various organic compounds, such as carbohydrates, fats and proteins, present in the green plants. Since virtually all other living organisms depend upon green plants for their energy, the efficiency of plants in any given area in capturing solar energy sets the upper limit to long-term energy flow and biological activity in the community.

The food manufactured by the green plants is utilized by themselves and also by herbivores. Animals feed repeatedly. Herbivores fall prey to some carnivorous animals. In this way one form of life supports the other form. Thus, food from one trophic level reaches to the other trophic level and in this way a chain is established. This is known as food chain.

A food chain may be defined as the transfer of energy and nutrients through a succession of organisms through repeated process of eating and being eaten. In food chain initial link is a green plant or producer which produces chemical energy available to consumers. For example, marsh grass is consumed by grasshopper, the grasshopper is consumed by a bird and that bird is consumed by hawk.

Thus, a food chain is formed which can be written as follows:

Marsh grass → grasshopper → bird → hawk

Food chain in any ecosystem runs directly in which green plants are eaten by herbivores, herbivores are eaten by carnivores and carnivores are eaten by top carnivores. Man forms the terrestrial links of many food chains.

Food chains are of three types

1. Grazing food chain
2. Parasitic food chain
3. Saprophytic or detritus food chain

Grazing Food Chain

The grazing food chain starts from green plants and from autotrophs it goes to herbivores (primary consumers) to primary carnivores (secondary consumers) and then to secondary carnivores (tertiary consumers) and so on. The gross production of a green plant in an ecosystem may meet three fates—it may be oxidized in respiration, it may be eaten by herbivorous animals and after the death and decay of producers it may be utilized by decomposers and converters and finally released into the environment. In herbivores the assimilated food can be stored as carbohydrates, proteins and fats, and transformed into much more complex organic molecules.

The energy for these transformations is supplied through respiration. As in autotrophs, the energy in herbivores also meets three routes respiration, decay of organic matter by microbes and consumption by the carnivores. Likewise, when the secondary carnivores or tertiary consumers eat primary carnivores, the total energy assimilated by primary carnivores or gross tertiary production follows the same course and its disposition into respiration, decay and further consumption by other carnivores is entirely similar to that of herbivores.

Thus, it is obvious that much of the energy flow in the grazing food chain can be described in terms of trophic levels as outlined below:

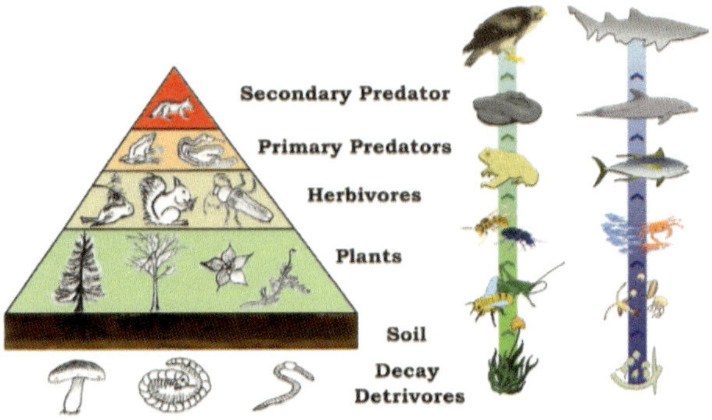

A schematic representation of grazing food chain showing input and losses of energy has been presented in Fig.

Parasitic Food Chain

It goes from large organisms to smaller ones without outright killing as in the case of predator.

Detritus Food Chain

The dead organic remains including metabolic wastes and exudates derived from grazing food chain are generally termed detritus. The energy contained in detritus is not lost in ecosystem as a whole, rather it serves as a source of energy for a group of organisms called detritivores that are separate from the grazing food chain. The food chain so formed is called detritus food chain.

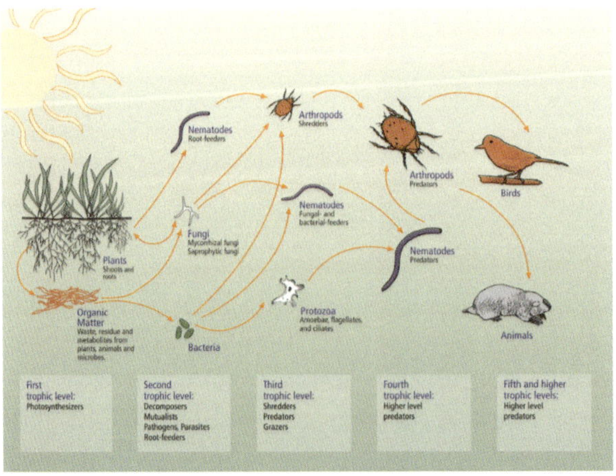

In some ecosystems more energy flows through the detritus food chain than through grazing food chain. In detritus food chain the energy flow remains as a continuous passage rather than as a stepwise flow between discrete entities. The organisms in the detritus food chain are many and include algae, fungi, bacteria, slime moulds, actinomycetes, protozoa, etc. Detritus organisms ingest pieces of partially decomposed organic matter, digest them partially and after extracting some of the chemical energy in the food to run their metabolism, excrete the remainder in the form of simpler organic molecules.

The waste from one organism can be immediately utilized by a second one which repeats the process. Gradually, the complex organic molecules present in the organic wastes or dead tissues are broken down to much simpler compounds, sometimes to carbon dioxide and water and all that are left are humus. In a normal environment the humus is quite stable and forms an essential part of the soil. Schematic representation of detritus food chain.

Food Web

Many food chains exist in an ecosystem, but as a matter of fact these food chains are not independent. In ecosystem, one organism does not depend wholly on another. The resources are shared specially at the beginning of the chain. The marsh plants are eaten by variety of insects, birds, mammals and fishes and some of the animals are eaten by several predators.

Similarly, in the food chain grass → mouse → snakes → owls, sometimes mice are not eaten by snakes but directly by owls. This type of interrelationship interlinks the individuals of the whole community. In this way, food chains become interlinked. A complex of interrelated food chains makes up a food web. Food web maintains the stability of the ecosystem. The greater the number of alternative pathways the more stable is the community of living things. Fig. illustrates a food web in ecosystem.

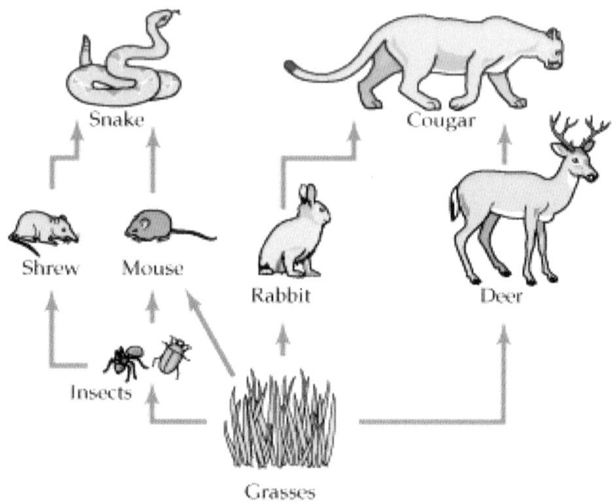

Food web in an ecosystem.

Ecological Pyramid

The trophic structure of an ecosystem can be indicated by means of ecological pyramid. At each step in the food chain a considerable fraction of the potential energy is lost as heat. As a result, organisms in each trophic level pass on lesser energy to the next trophic level than they actually receive. This limits the number of steps in any food chain to 4 or 5. Longer the food chain the lesser energy is available for final members. Because of this tapering off of available energy in the food chain a pyramid is formed that is known as ecological pyramid. The higher the steps in the ecological pyramid the lower will be the number of individuals and the larger their size.

The idea of ecological pyramids was advanced by C.E. Elton. There are different types of ecological pyramids. In each ecological pyramid, producer level forms the base and successive levels make up the apex. Three types of pyramidal relations may be found among the organisms at different levels in the ecosystem.

These are as follows:

1. Pyramid of numbers,
2. Pyramid of biomass (biomass is the weight of living organisms), and
3. Pyramid of energy.

Pyramid of Numbers

It depicts the numbers of individuals in producers and in different orders of consumers in an ecosystem. The base of pyramid is represented by producers which are the most abundant. In the successive levels of consumers, the number of organisms goes on decreasing rapidly until there are a few carnivores.

The pyramid of numbers of an ecosystem indicates that the producers are ingested in large numbers by smaller numbers of primary consumers. These primary consumers are eaten by relatively smaller number of secondary consumers and these secondary consumers, in turn, are consumed by only a few tertiary consumers.

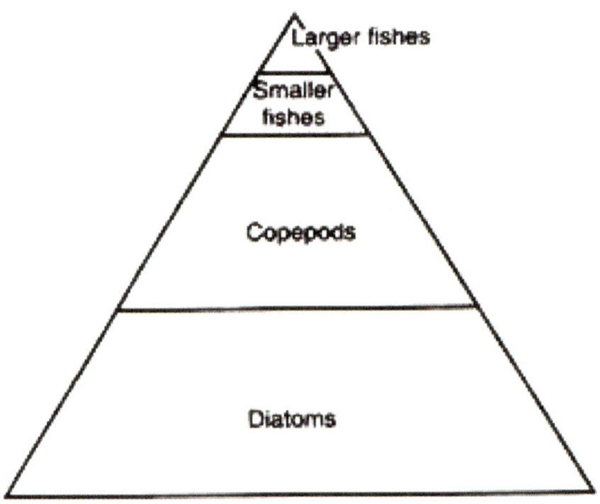

Figure: A pyramid of numbers of a lake ecosystem.

This type of pyramid is best presented by taking an example of Lake Ecosystem. In this type of pyramid the base trophic level is occupied by producer elements—algae, diatoms and other hydrophytes which are most abundant. At the second trophic level come the herbivores or zooplanktons which are lesser in number than producers.

The third trophic level is occupied by carnivores which are still smaller in number than the herbivores and the top is occupied by a few top carnivores. Thus, in the ecological pyramid of numbers there is a relative reduction in number of organisms and an increase in the size of body from base to apex of the pyramid. In parasitic food chain starting from tree, the pyramid of numbers will be inverted.

Figure: Pyramids of number. (a) Upright pyramids of numbers in a grassland and cultivated field (b) pyramids of number inverted of discuased tree (parasitic ecosystem)

Pyramid of Biomass of Organisms

The living weights or biomass of the members of the food chain present at any one time form the pyramid of biomass of organisms. This indicates, by weight or other means of measuring materials, the total bulk of organisms or fixed energy present at one time. Pyramid of biomass indicates the decrease of biomass in each tropic level from base to apex, e.g., total biomass of producers is more than the total biomass of the herbivores.

Likewise, the total biomass of secondary consumers will be lesser than that of herbivores and so on. Since some energy and material are lost in each successive link, the total mass supported at each level is limited by the rate at which the energy is being stored below. This usually gives sloping pyramid for most of the communities in terrestrial and shallow water ecosystems. The pyramid of biomass in a pond ecosystem will be inverted as shown in fig.

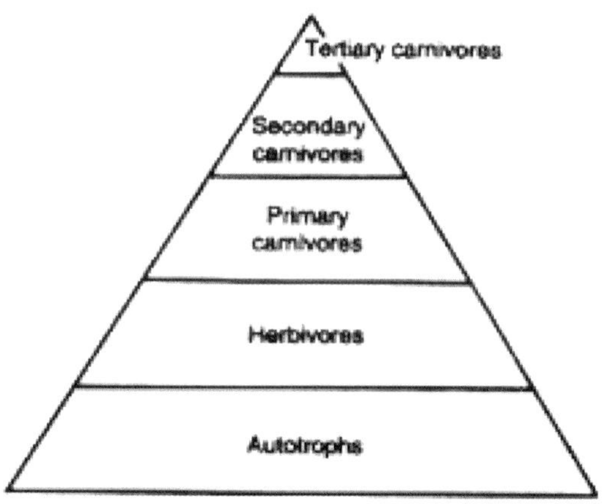

Figure: A pyramid of biomass.

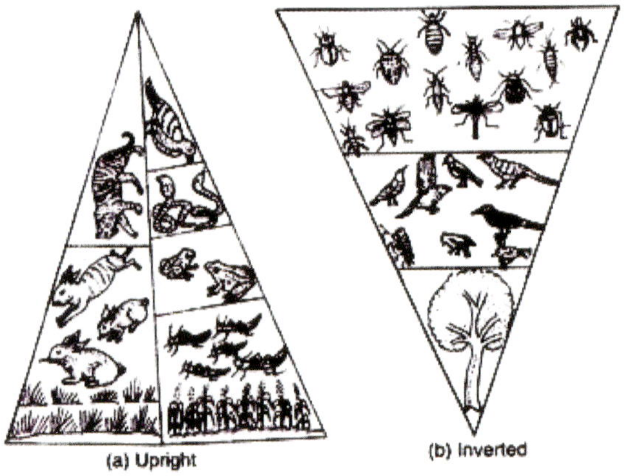

Figure: Pyramids of biomass (a) A grassland ecosystems showing upright-triangular (b) Inverted pyramid of biomass of an aquatic ecosystem.

Pyramid of Energy:

This depicts not only the amount of total energy utilized by the organisms at each trophic level of food chain but more important, the actual role of various organisms in transfer of energy. At the producer level the total energy will be much greater than the energy at the successive higher trophic level.

Some producer organisms may have small biomass but the total energy they assimilate and pass on to consumers may be greater than that of organisms with much larger biomass. Higher trophic levels are more efficient in energy utilization but much heat is lost in energy transfer. Energy loss by respiration also progressively increases from lower to higher trophic states.

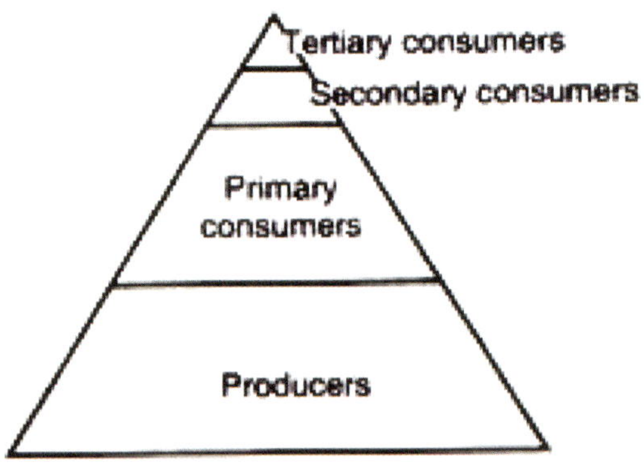

Figure: A pyramid of energy.

In the energy flow process, two things become obvious. Firstly there is one way along which energy moves i.e. unidirectional flow of energy. Energy comes in the ecosystem from outside source i.e. sun. The energy captured by autotrophs does not go back to the sun, the energy that passes from autotrophs to herbivores does not revert back and as it moves progressively through the various trophic levels, it is no longer available to the previous levels.

Thus due to unidirectional flow of energy, the system would collapse if the supply from primary source, the sun is cut off. Secondly, there occurs a progressive decrease in energy level at each trophic level which is accounted largely by the energy dissipated as heat in metabolic activities.

Productivity

The relationship between the amount of energy accumulated and the amount of energy utilized within one trophic level of food chain has an important bearing on how much energy from one trophic level passes on to the next trophic level in the food chain. The ratio of output of energy to input of energy is referred to as ecological efficiency.

Different kinds of efficiencies can be measured by the following parameters:

(i) Ingestion which indicates the quantity of food or energy taken by trophic level. This is also called exploitation efficiency.

(ii) Assimilation indicates the amount of food absorbed and fixed into energy rich organic substances which are stored or combined with other molecules to build complex molecules such as proteins, fats etc.

(iii) Respiration which indicates the energy lost in metabolism.

Primary Productivity

The fraction of fixed energy a trophic level passes on to the next trophic level is called production. Green plants fix solar energy and accumulate it in organic forms as chemical energy. Since it is the first and basic form of energy storage, the rate at which the energy accumulates in the green plants or producers is known as primary productivity.

Primary productivity is the rate at which energy is bound or organic material is created by photosynthesis per unit area of earth's surface per unit time. It is most often expressed as energy in calories/cm^2/yr or dry organic matter in g/m^2/yr (g/m^2 x 8.92 = lb/acre). The amount of organic matter present at a given time per unit area is called standing crop or biomass and as such productivity, which is a rate, is quite different from biomass or standing crop.

The standing crop is usually expressed as dry weight in g/m^2 or kg/m^2 or t/ha (metric tons) or 10^6g/hectare. Primary productivity is the result of photosynthesis by green plants including algae of different colours. Bacterial photosynthesis or chemosynthesis, although of small significance may also contribute to primary productivity. The total solar energy trapped in the food material by photosynthesis is referred to as gross primary productivity (G.P.P.).

A good fraction of gross primary production is utilized in respiration of green plants. The amount of energy bound in organic matter per unit area and time that is left after respiration in plants is net primary production (N.P.P.) or plant growth. Only the net primary productivity is available for harvest by man and other animals. Net productivity of energy = gross productivity—energy lost in respiration.

Secondary Productivity

The rates at which the heterotrophic organisms resynthesize the energy-yielding substances is termed as secondary productivity. Secondary productivities are the productivities of animals and saprobes in communities. The amount of energy stored in the tissues of consumers or heterotrophs is termed as net secondary production and the total plant material ingested by herbivores is grass secondary production. Total plant material ingested by herbivores minus the materials lost as faeces is equal to Ingested Secondary Production.

Environmental factors affecting the production processes in an ecosystem are as follows:

1. Solar radiation and Temperature
2. Moisture. Leaf water potential, soil moisture and precipitation fluctuation and transpiration.
3. Mineral nutrition. Uptake of minerals from the soil, rhizosphere effects, fire effects, salinity, heavy metals, nitrogen metabolism.
4. Biotic activities, Grazing, above ground herbivores, below ground herbivores, predators and parasites, diseases of primary producers.
5. Impact of human population. Pollutions of different sorts, ionizing radiations like atomic explosions, etc.

There are three fundamental concepts of productivity

1. Standing crop;
2. Materials removed;
3. Production rate.

Standing Crop

It is the abundance of the organisms existing in the area at any one time. It may be expressed in terms of number of individuals, as biomass of organisms, as energy content or in some other suitable terms. Measurement of standing crop reveals the concentration of individuals in the various populations of ecosystem.

The Materials Removed

The second concept of productivity is the materials removed from the area per unit time. It includes the yield to man, organisms removed from the ecosystem by migration, and the material withdrawn as organic deposit.

The Production Rate:

The third concept of productivity is the production rate. It is the rate at which the growth processes are going forward within the area. The amount of material formed by each link in the food chain per unit of time per unit area or volume is the production rate.

All the three major groups of organisms—producers, consumers and reducers are the functional kingdoms of natural communities. The three represent major directions of evolution and are characterised by different modes of nutrition. Plants feed primarily by photosynthesis, animals feed primarily by ingesting food that is digested and absorbed in the alimentary canal and the saprobes feed by absorption and have need for an extensive surface of absorption. The principal kinds of organisms among saprobes are the unicellular bacteria, yeasts, chytrids or lower fungi and higher fungi with mycelial bodies.

In terrestrial communities as much as 90% of net primary production remains un-harvested and are utilized as dead tissue by saprobes and soil animals. The saprobes have a larger and more essential role than animals in degrading dead organic matter to inorganic forms and in such ecosystems, secondary production by reducers (decomposers) should exceed that by consumers, though the former is even more difficult to measure than the latter.

Biomass of decomposers with their microscopic cells and filaments embedded in food sources is also difficult to measure and that is small in relation to their productivity and significance for the ecosystem. Small masses of reducers degrade and transform larger masses of organic matter to inorganic remnants. In so doing decomposers disperse back to the environment the energy of photosynthesis accumulated in the organic compounds that are decomposed.

Thus they have a major role in the energy flow of ecosystems. A community or ecosystem, like an organism, is an open energy system. The continuous intake of energy in photosynthesis replaces the energy dissipated to environment by respiration and biological activity and the system does not run-down through the loss of free energy to maximum entropy.

If the amount of energy entrapped is greater than the energy dissipated, the pool of biologically useful energy of organic bonds increases. This results in increase of community biomass and consequently the community grows; such is the case in succession. If energy intake is lesser than energy dissipation, the community biomass will decrease and it must, in some sense, retrogress. If energy intake and loss are in balance, the pool of organic energy is in steady state; such is the case in climax communities.

Three aspects of this steady state may be recognized

(i) The steady state of population of climax communities in which equal birth and death rates in population keep the number of individuals relatively constant,

(ii) The steady state of energy flow,

(iii) The steady state of the matter of community, where addition of material by photosynthesis and organic synthesis is balanced by loss of material through respiration and decomposition.

Methods of Measuring Primary Production

There are several parameters for measuring primary production and the methods of measuring primary production are based on those parameters.

The methods are discussed here as under

Harvest Method

It involves removal of vegetation periodically and weighing the material. For measuring above ground production, the above ground plant parts are clipped at ground level, dried to constant weight at 80°C and weighed. The dry weight in $g/m^2/year$ gives the ground production. Below ground production is estimated by using frequent core sampling technique of Dahlman and

Kucera. It is expressed in terms of weight in gm per unit area per year. In terms of energy one gm dry weight of plant material contains 4 to 5 kcal.

The limitations of harvest method are as follows:

(i) The amount of plant material consumed by herbivores and the food oxidized during respiration process of the plants is not accounted.

(ii) Root biomass is neglected.

(iii) Photosynthetic trans located to underground parts of plants are not known.

In spite of these limitations the method is used all over for measuring net assimilation rate (NAR) and relative growth rate (RGR).

Carbon Dioxide Assimilation Method:

Utilization of CO_2 in photosynthesis or its liberation during respiration is measured by infrared gas analysis or by passing the gas through Baryta water $Ba(OH)_2$ and titrating the same. The CO_2 removed from incoming gas chamber is taken to be synthesized into organic matter by the green plants. Performing the experiment in light and dark chambers the net and gross production can be measured.

In the lighted chamber photosynthesis and respiration take place simultaneously and the CO_2 coming out from the chamber is the unused gas of the atmosphere plus gas from the respiration of plant parts. In the dark chamber all CO_2 is due to respiration.

Net production = Gross production—Respiration

Oxygen Production Method:

In the aquatic vegetation CO_2 gas analysis method is not used but oxygen evolution method is generally used. The light and dark bottle technique is employed for measuring primary production of aquatic plant. In this method two bottles, one transparent and the other opaque are filled with water at a given depth of lake, closed, maintained at that depth for some time and then brought to laboratory for determination of oxygen content in the water. The decrease of oxygen in dark bottle is due to respiratory activity while increase of O_2 in light bottle is due to photosynthesis. The total increase of O_2 in light bottle plus the amount of O_2 decreased in dark bottle express gross productivity (O_2 value multiplied by 0.375 gives an equivalent of carbon assimilation). Recently, oxygen electrodes have been used for estimating oxygen content in water.

Chlorophyll Method

Gesner (1949) pointed out that the amount of chlorophyll/m^2 is almost limited to a narrow range of 0.1 to 3.0 gm regardless of the age of individuals or the species present therein. There is direct correlation between the amount of chlorophyll and dry matter production in different types of communities with varying light conditions.

The relation of total amount of chlorophyll to the photosynthetic rate is referred to as assimilation ratio or rate of production/gm chlorophyll. Total chlorophyll per unit area is greater in land plants as compared to that in aquatic plants. In marine ecosystem the rate of carbon assimilation is 3.7 g/

hr/g of chlorophyll. The relationship between area based chlorophyll and dry matter production in terrestrial ecosystems has been worked out by Japanese ecologists Argua and Monsi (1963).

Other Methods

Several ecologists have evolved correlation coefficients for evaluating biomass and productivity in forest trees by measuring their diameter at breast height (DBH), height, canopy cover, etc.

Methods of establishing regression are as below

(i) Diameter of trees in sample quadrats is measured at breast height and the height repeated is determined for each tree.

(ii) Different diameter and height classes are determined for each species.

(iii) A set of sample trees are cut and subjected to a detailed analysis for dry weight of stems, twigs, leaves and roots.

(iv) Regression values are computed for the sets of trees belonging to each girth class, relating the biomass of each fraction to the diameter at breast height.

(v) The regression values are used to compute the probable biomass and Production each tree in the sample area. These values for each species when pooled give biomass and production rate of trees per unit area in the forest. Age of the trees markedly influences the annual net production.

Primary Consumers

Primary consumers make up the second trophic level as they feed on primary producers. In other words, primary consumers are organisms that take up food directly from plants.

Primary consumers are normally herbivores. They are also referred to as heterotrophs as they don't produce their own food. They largely rely on autotrophs, which are plants that produce their food through photosynthesis.

These organisms use several other feeding strategies. Algivores, for instance, feed on photosynthetic algae. Nectavores consume plant nectar, frugivore consumes fruiting bodies of plants, granivores feed on seeds and grains, folivores consume leaf material, while fungivores eat heterotrophic fungi such as mushrooms.

Primary consumers are found in all biomes and cut across a wide range of niches.

Examples of Primary Consumers

Primary consumers can range from microscopic organisms like zooplankton to large creatures like elephants. Here are some examples.

1. Ruminants like Giraffes and Cows

Primary herbivorous consumers such as cows, goats, zebras, giraffes are primary consumers. They consume plant material such as grass, branches, and roots.

Cell walls of plants have cellulose that is difficult to break. In light of this, ruminants have an adaptive system that enables them to obtain nutrients through fermentation and digestion. The process takes place in four specialized chambers of their stomach.

From a ruminant's mouth, chewed plant material is directed to the first digestive chamber, the rumen and the reticulum. In this chamber, the food is mixed with saliva. It is then separated into solid mass and liquid form.

The solid mass is returned to the mouth, where it is chewed to reduce it into as small food particles as possible. These small food particles are then returned into the digestive chambers. In the chambers, cellular fibroids are broken by protozoa, fungus, and bacteria.

The complexity of the ruminants' stomach serves to show the challenges that large animals experiencing when it comes to extracting nutrients from plant carbohydrates. Nonetheless, the volatility of the proteins and fatty acids that are generated from this system form an integral component of the human diet.

2. Herbivorous Birds

While some bird species are carnivorous or omnivorous, many birds eat only seeds, cherries, and fruits. This qualifies them to be herbivorous, and hence primary consumers.

Birds that base their diet on plant material normally have morphologically adapted beaks. With such beaks, they can easily exploit their food source.

While some bird species are carnivorous or omnivorous, many birds eat only seeds, cherries, and fruits. This qualifies them to be herbivorous, and hence primary consumers.

Birds that base their diet on plant material normally have morphologically adapted beaks. With such beaks, they can easily exploit their food source.

Parrots, parakeets, and toucans, are endowed with exceptionally strong beaks with which they can effortlessly crack nuts. The beaks also act as a stabilizing tool for climbing large trees and reach the fruits at the highest points of the tree.

Conversely, hummingbirds and other similar species normally have very small, elongated and pointed beaks. With such beaks, they can reach nectar deep inside the flowers.

In light of the fact that they largely feed on grains and seeds, many canaries, finches, and parrots have a short, hard, and pointed beak. Such beaks enable them to pick the grains with great precision.

3. Zooplankton

Zooplankton are microscopic organisms that live as accumulated organisms in the marine ecosystem. Zooplankton includes protozoa and metazoans that are in their junior stage. These metazoans include crustaceans and mollusks.

Zooplankton organisms are largely heterotrophic. Namely, they obtain their nutrients from the coal generated through photosynthesis. This process is facilitated by the primary producers of the marine ecosystem, which transform organic coal into usable energy.

Moreover, zooplankton are virtually the primary producers of food through its filtration. In this mode of feeding, water is passed through specialized filters where phytoplankton present in the water is filtered and digested.

There are numerous other primary consumers on earth. Most of them are much smaller and with little impact on the ecosystem. The squirrel, mouse, and the beetle are examples of such primary consumers.

It is also worth noting that some primary consumers may also fit in other trophic levels as their diet varies. For example, when a bear feeds on berries, it qualifies to be a primary consumer. When the bear consumes salmon, it is functioning as a tertiary consumer. Note that salmon is a secondary consumer because it eats herring, which feeds on zooplankton that eats phytoplankton. Phytoplankton, on the other hand, make their own food using the sun.

Secondary Consumers

Secondary consumers are a category of living organisms that feed on primary consumers to get energy. The secondary consumers are placed on the upper levels of the food chain. Primary producers make their own food in the typical food chain in a process known as photosynthesis whereas the primary consumers (herbivores) eat the primary producers, which consist majorly of plants. Secondary consumers occupy the third and fourth trophic levels in the food chain, and there are two groups of these living organisms, omnivores, and carnivores. The omnivore secondary consumers eat both plants and animals while the carnivores purely feed on animals.

The Importance of Secondary Consumers

Since the food chain comprises of different levels, each level has its significance in the system. The secondary consumers play specific vital roles in the food chain by taking part in controlling the population of primary consumers in the ecosystem. They control the population by feeding on the primary consumers, ensuring that they cannot exceed the number that the ecosystem can withstand. Moreover, secondary consumers are a source of energy to the tertiary consumers.

Examples of Secondary Consumers

The food chain provides a number of various examples of secondary consumers. First, there are the carnivores who eat flesh from other animals only. The carnivores may range from small insects to large predators. Interestingly, most small carnivores may eat other animals which have a larger body size than theirs. Examples of such carnivores include spiders, snakes, and seals. Examples of large carnivores are lions, leopards, crocodiles, and wolves. An eagle is an example of a predator bird that falls under the category of secondary consumers.

Tertiary Consumers

A tertiary consumer is a fourth trophic level after producers, primary consumers, and secondary consumers.

Tertiary consumers eat primary and secondary consumers as their main source of food. These organisms are sometimes referred to as apex predators as they are normally at the top of food chains, feeding on both primary and secondary consumers.

Tertiary consumers can be carnivores or omnivores. Their diet can comprise only meat or include plants as well. A hawk, for example, can feed on primary consumers such as birds, as well as secondary consumers such as snakes.

Examples of Tertiary Consumers

Many organisms, both in terrestrial and aquatic ecosystems, are considered to be tertiary consumers. Here are some of them:

Humans

Humans are largely omnivorous. They eat both plants and animals. Moreover, they have a wide-ranging diet and hence consume foods from virtually all trophic levels. We even consume decomposers such as mushrooms.

If you choose to be a vegetarian, you would be considered to be a primary consumer, thanks to the fact that you only consume plant materials. But if you also choose to eat chicken that only feeds on grain, you would fit in the secondary consumer category.

However, if the chicken also feeds on insects, then you would be a tertiary consumer. In light of the fact that humans can kill any animal using weapons, they are of considered as apex predators.

Big Cats like Lions and Tigers

Lions, tigers, leopards and other big cats are classed as tertiary consumers. Except for the leopard, the big cats are also apex predators. While the cats have no predators in their natural habitat, leopard is sometimes predated by lions and tigers.

The big cats exhibit various characteristics of apex predators, including:

- Large teeth;
- Strong Jaws;
- Large Claws;
- Forward facing eyes for tracking prey;
- Strong muscles;
- Ability to sprint at great speed.

Big cats feed on animals from all trophic levels below them. These animals include herbivores that stay in herds such as zebras, wildebeest, and buffalo. Secondary consumers such as hyenas and foxes are also part of their menu.

Furthermore, the big cats sometimes feed on large animals such as crocodiles. They prey on crocodiles when on land because crocodiles have an upper hand when in the water and may attack the big cats instead.

Polar Bear

The polar bear is the leading predator in the Arctic. It is a tertiary predator in light of the fact that it kills fish, seals, and penguins.

However, if a polar bear consumes a seal that has eaten a penguin that has eaten a fish, it may not be considered a tertiary predator. This is because several trophic levels have been passed in the process. Nonetheless, this complexity is normally ignored and polar bears are classed as tertiary and apex predators.

Secretary Bird

Secretary birds are renowned snake hunters. They hunt, kill, and consume various reptiles and small mammals. In fact, venomous snakes such as adders and cobras are their favorite meal.

Adult secretary birds are endowed with exceptionally strong legs, which can break your bones with a single strike. This powerful weapon makes this bird virtually invincible.

Secretary birds are apex predators, but their chicks could be eaten.

Crocodiles

Crocodiles rule the waters. They are exceptionally powerful when in the water. They are one of the animals with the strongest jaws in the animal kingdom.

That means they boast one of the most powerful bites. Even the big cats such as tigers and lion can't beat crocodiles if they enter the water.

Pythons and Boas

Pythons can effortlessly kill wildcats and foxes. But they would have a hard time beating predators such as big cats and adult crocodiles.

These giant snakes can also kill young big cats and crocodiles. Anacondas normally prey on caimans, which are much smaller than crocodiles.

Other Examples of Marine Tertiary Consumers

Marine ecosystems also have a substantial share of tertiary consumers. Phytoplankton, the primary producers in the aquatic environment are normally fed upon by microscopic organisms knows as zooplankton. Hence, the numerous animals that consume the zooplankton are secondary consumers.

These animals include fish, crustaceans, and jellyfish. Tertiary consumers, on the other hand, include:

- Large fish such as barracuda, tuna, and groupers;
- Seals and sea lions;
- Dolphins;
- Turtles;

- Sharks;
- Moray eels;
- Whales.

Some of these organisms are apex predators, such as orca whales and the great white or tiger sharks. Moreover, many seabirds such as shearwaters, penguins, and gulls are tertiary consumers.

In freshwater ecosystems, predatory fish like spike eat smaller fish and other secondary consumers such as snakes, birds, frogs, and small mammals.

Functions of Tertiary Consumers

The energy present within organisms in an ecosystem is normally conveyed through a food chain or food web. Every organism in a food chain belongs to a specific position known as a trophic level. Animals eat other animals at lower trophic levels and are consumed by those at higher trophic levels.

At the top trophic level, we normally have tertiary consumers, which are not predated by any other animals. And hence the name Apex Predators.

However, when tertiary consumers die, their bodies are consumed by scavengers and decomposers. In some food chains, an apex predator exists above the tertiary consumer.

Nonetheless, energy is expended and is lost as heat as it is passed through every trophic level. This leads to a limited energy in the higher trophic levels. Therefore, it is common to only have four trophic levels. And the tertiary consumer is left to play the ecological role of the apex predator.

Organisms in the topmost trophic levels play an integral role in ecosystems. Besides controlling the population of species at lower trophic levels, they also alter their behavior.

When the population of animals in the lower trophic levels are limited, it eases the pressure of grazing or predation on the trophic levels below them. When this happens, the ecosystem dynamics remain balanced.

Let's use foxes, hares, and hawks to illustrate this phenomenon. When a population of foxes becomes too large, it would put pressure on the populations of hares. By predating on the foxes, a hawk regulates their populations and hence reduce the number of hares that are eaten by the foxes.

That phenomenon is known as a trophic cascade.

Decomposers

A decomposer is an organism that decomposes, or breaks down, organic material such as the remains of dead organisms. Decomposers include bacteria and fungi. These organisms carry out the process of decomposition, which all living organisms undergo after death. Decomposition is an important process because it allows organic material to be recycled in an ecosystem.

Mushrooms, such as those in the image above, are a type of fungus and play a role in decomposition.

Function of Decomposers

Decomposers play an important role in every ecosystem. Without decomposers, dead organisms would not be broken down and recycled into other living matter. The reason decomposers decompose, however, is simply because they need to survive. Decomposers are heterotrophic, which means they get their energy from ingesting organic material. A dead organism provides nutrients for decomposers like bacteria and fungi to use in order to grow and reproduce, propagating their own species. The side effect of this basic need to survive is that organic material and nutrients are cycled throughout the ecosystem as other organisms consume the bacteria and fungi.

Decomposers and Detritivores

Some organisms perform a similar function as decomposers, and are sometimes called decomposers, but are technically detritivores. The difference lies in the way decomposers and detritivores break down organic material. Detritivores must digest organic material within their bodies in order to break it down and gain nutrients from it. Decomposers do not need to digest organic material internally in order to break it down; instead, they can break down matter through biochemical reactions. Organisms that are detritivores include invertebrates such as earthworms, woodlice, sea stars, slugs, and fiddler crabs.

Decomposers and Scavengers

Scavengers are the first to arrive at a dead organism's remains, and they directly eat the dead plant and animal material. Once scavengers are done with the remains, decomposers and detritivores take over and consume the parts that the scavengers have left behind. Many predators will scavenge on occasion; examples of these sometime scavengers include lions, jackals, wolves, raccoons, and opossums. Vultures are obligate scavengers, meaning that scavenging is how they obtain all of their food. They are the only members of the animal kingdom that have to scavenge in order to eat.

Stages of Decomposition

When an organism dies and decomposers do the work of decomposition, the organism's remains go through five stages of decomposition: fresh, bloat, active decay, advanced decay, and dry/remains. There are two main processes that occur in a decomposing organism: autolysis and putrefaction. Autolysis is when cellular enzymes in the dead organism's own body break down cells and tissues, while putrefaction is when microbes grow and reproduce throughout the body after death. Here is a brief summary of the five stages.

Fresh

This stage begins as soon as an organism's heart stops beating. With no more oxygen coming into the body and a build-up of carbon dioxide, autolysis begins to occur. Putrefaction also begins to occur.

Bloat

Due to putrefaction, a build-up of gases occurs and the organism's remains appear bloated in what is known as the bloat stage. Some gases and fluids are purged from the body.

Active Decay

The remains lose mass, and liquefaction and disintegration of tissues begins to occur. Bacteria produce chemicals such as ammonia, hydrogen sulfide, and methane, which cause strong odors.

Advanced Decay

The organism has lost a lot of mass, so there is not much left to be decomposed. If the organism is on or in soil, the surrounding soil will show an increase in nitrogen, an important nutrient for plants.

Dry/Remains

In this stage, only dry skin, cartilage, and bones are left. Plant growth may occur around the remains because of the increased nutrient levels in the soil. Eventually, only the bones of the organism will remain.

Examples of Decomposers

Bacteria

Bacteria are microscopic, unicellular organisms found nearly everywhere on Earth, including inside the human body. When an organism dies, it provides many nutrients for bacteria to grow and reproduce, and they become numerous in the process of putrefaction during decomposition. Bacteria are themselves a cause of sickness and even death when they infect organisms. Serious and often fatal diseases such as typhus, tuberculosis, and cholera are caused by bacterial infections. Bacteria that kill their hosts end up inadvertently providing nutrients for other bacteria during decomposition.

Fungi

Fungi are the main decomposers in many environments. Some examples of fungi are yeasts, molds, and mushrooms. Fungi have hyphae, which are branching filaments, and these hyphae are able to enter organic matter, making fungi effective decomposers. Wood-decay fungi have specific enzymes that digest compounds in wood, and are the main decomposers in forests. In fact, wood-decay fungi are the only producers of these enzymes, so they play a very important role in decomposition.

Biodiversity

Biodiversity or Biological diversity is a term that describes the variety of living beings on earth. In short, it is described as degree of variation of life. Biological diversity encompasses microorganism, plants, animals and ecosystems such as coral reefs, forests, rainforests, deserts etc.

Biodiversity also refers to the number, or abundance of different species living within a particular region. It represents the wealth of biological resources available to us. It's all about the sustaining the natural area made up of community of plants, animals, and other living things that is begin reduced at a steady rate as we plan human activities that is being reduced by habitat destruction.

The United Nations designated 2011–2020 as the United Nations Decade on Biodiversity. In biodiversity, each species, no matter how big or small has an important role to play in ecosystem. Various plant and animal species depend on each other for what each offers and these diverse species ensures natural sustainability for all life forms. A healthy and solid biodiversity can recover itself from variety of disasters.

Biodiversity has three essential elements:

- Genetic diversity,
- Eco system diversity and,
- Species diversity.

Recently a new aspect has also been added- 'molecular diversity'.

Biodiversity is unevenly distributed. It varies globally and within regions. The various factors that influence biodiversity include -temperature, altitude, precipitation, soils and their relation with other species. For instance, ocean biodiversity is 25 times lesser than terrestrial diversity. Biodiversity also increases its form as it moves from the poles towards the tropics.

Biodiversity is the result of 3.5 billion years of evolution. It has been subject to periods of extinction. The latest and most destructive stage of extinction is Holocene extinction, which has occurred due to the impact of human beings on the environment.

Importance of Biodiversity

Biodiversity has a number of functions on the Earth. These are as follows:

- Maintaining balance of the ecosystem: Recycling and storage of nutrients, combating pollution, and stabilizing climate, protecting water resources, forming and protecting soil and maintaining ecobalance.
- Provision of biological resources: Provision of medicines and pharmaceuticals, food for the human population and animals, ornamental plants, wood products, breeding stock and diversity of species, ecosystems and genes.
- Social benefits: Recreation and tourism, cultural value and education and research.

The role of biodiversity in the following areas will help make clear the importance of biodiversity in human life:

- Biodiversity and food: 80% of human food supply comes from 20 kinds of plants. But humans use 40,000 species for food, clothing and shelter. Biodiversity provides for variety of foods for the planet.
- Biodiversity and human health: The shortage of drinking water is expected to create a major global crisis. Biodiversity also plays an important role in drug discovery and medicinal resources. Medicines from nature account for usage by 80% of the world's population.
- Biodiversity and industry: Biological sources provide many industrial materials. These include fiber, oil, dyes, rubber, water, timber, paper and food.
- Biodiversity and culture: Biodiversity enhances recreational activities like bird watching, fishing, trekking etc. It inspires musicians and artists.

Reason for Loss of Biodiversity

The earth's biodiversity is in grave danger. In the present era, human beings are the most dangerous cause of destruction of the earth's biodiversity. In 2006, the terms threatened, endangered or rare were used to describe the status of many species. The "evil quartet" identified by Jared Diamond is overkill, habitat destruction, secondary extinctions and introduced species. Factors identified by Edward Wilson are described by the acronym- HIPPO standing for habitat destruction, climate change, invasive species, pollution, human overpopulation and over-harvesting.

Habitat destruction is a major cause for biodiversity loss. Habitat loss is caused by deforestation, overpopulation, pollution and global warming. Species which are physically large and those living in forests or oceans are more affected by habitat reduction.

Some expert's estimate that around 30% of all species on earth will be extinct by 2050. According to the International Union for Conservation of Nature (IUCN), globally about one third of all known species are threatened with extinction. Even it is estimated that 25% of all mammals will be extinct within 20 years.

Even if a small element of an ecosystem breaks down, the whole system's balance is threatened. Fresh water ecosystems are nowadays the most threatened ecosystems. Invasive species refer to those that would normally remain constrained from an ecosystem because of the presence of natural barriers. Since these barriers are no longer existing, invasive species invade the ecosystem, destroying native species. Human activities have been the major cause for encouraging invasive species.

Species can also be threatened by genetic pollution- uncontrolled hybridization and gene swamping. For instance, abundant species can interbreed with rare species thus causing swamping of the gene pool. Over exploitation is caused by activities such as over fishing, over hunting, excessive logging and illegal trade of wildlife. Over 25% of global fisheries are being overfished at unsustainable levels.

Global warming is also becoming a major cause for loss of biodiversity. For example if the present rate of global warming continues, coral reefs which are biodiversity hotspots will disappear in 20-40 years. 10% of all species might become extinct by 2015, if global warming continues.

Thus we can see that biodiversity which is crucial for the well being of life on earth, is coming under the threat of many factors related to human activities. There is an urgent need to take action to protect the magnificent biodiversity of our planet. We must create economic policies in order to maintain the Earth's biodiversity and take appropriate measures to protect habitats and species.

Ecosystem Services

Ecosystem services are outputs, conditions, or processes of natural systems that directly or indirectly benefit humans or enhance social welfare. Ecosystem services can benefit people in many ways, either directly or as inputs into the production of other goods and services. For example, the pollination of crops provided by bees and other organisms contributes to food production and is thus considered an ecosystem service. Another example is the attenuation of flooding in residential areas provided by riparian buffers and wetlands.

Apple pollination: A honeybee pollinating apple blossoms. The pollination of food crops by bees and other insects is an example of an ecosystem service that directly contributes to social welfare.

Because ecosystem services are not usually bought and sold directly in markets, market activities do not fully reflect the benefits provided by those services. Unregulated markets thus promote excessive depletion of natural capital (e.g., the biotic and abiotic components of ecosystems) and ecosystem services. The United Nations Millennium Ecosystem Assessment (2005), which evaluated the consequences of ecosystem change, concluded that humans have degraded the ability of Earth's ecosystems to support social welfare. In response, ecosystem services analyses promote policy decisions that recognize the full range of benefits and costs associated with actions that affect those services. Most formal evaluations of ecosystem services examine the consequences of changes to specific services in certain geographical areas for particular beneficiary groups; very few analyses approach ecosystem services from universal perspectives (e.g., all the services provided by wetlands across the planet).

Identification, Quantification, and Valuation

Two criteria distinguish ecosystem services from other ecosystem conditions or processes. First, an ecosystem service must be linked to an identifiable set of human beneficiaries. The service can be an aspect or consequence of an ecological condition and can directly or indirectly benefit or profit the beneficiaries. Second, physical and institutional access constraints must not prevent people from realizing those benefits. For example, increases in fish abundance can enhance the welfare of those engaged in commercial or recreational fishing, but only if those increases occur in areas where fishing occurs. Conditions or processes of ecosystems that cannot be linked to the welfare of identifiable beneficiary groups are not ecosystem services. For example, changes in fish abundance in areas not used by humans and that have no direct or indirect effect on human benefits are not ecosystem services.

The first step in most ecosystem services assessments is to identify the services to be evaluated. This requires an understanding of how beneficiaries are affected directly or indirectly by changes in ecological conditions and processes and can be assisted by the development of a conceptual causal chain. Such a chain clarifies the links between human actions and ecological effects and the subsequent changes to ecosystem services and the associated human benefits. For example, a causal chain related to mechanical forest thinning would first identify the primary effects of thinning on ecological conditions such as forest structure. Those changes are then linked to impacts on ecosystem outputs, conditions, or processes that affect the welfare of identifiable beneficiary groups; those are the ecosystem services. Forest thinning, for instance, could change forest structure in such a way that there is a reduction in fire risk to populated areas. Causal chains often include multiple steps between the original human action and the effects on final ecosystem services. One of the challenges of ecosystem services analysis is identifying the many ways that ecosystem services affect different beneficiary groups.

A common second step in an ecosystem service assessment is to quantify one or more of the services identified in the causal chain. Quantification requires an understanding of the measures of quantity, quality, or other biophysical features that are most directly relevant to human welfare, followed by the use of appropriate methods to obtain those measures. Some ecosystem services are easily quantified, such as the quantity of timber produced in a given forest area. Others are more difficult to measure, either because of limitations in biophysical models or data (e.g., fish produced by tidal wetlands) or because the service is inherently difficult to quantify (e.g., aesthetic features of natural landscapes).

The third step in an ecosystem service assessment is to determine the consequences for social welfare. This is often conducted by using formal economic valuation methods grounded in neoclassical economic theory, although noneconomic techniques may also be applied. Economic valuation quantifies the value of an ecosystem service in commensurable (typically monetary) units; that value is well-defined only for a specific quantity of the ecosystem service, measured from a particular baseline. Values calculated for services at a biome or planetary scale (for which changes and baselines are often poorly defined) are generally considered invalid—or at least highly inaccurate—by economists. The accuracy of valuation also depends on the ability of an analysis to account for factors that influence the contribution of ecosystem services to welfare. For example, all else equal, per-unit values often increase as a service becomes more scarce (reflecting diminishing marginal utility) and are frequently affected by spatial factors such as the distance of beneficiaries from the service.

Valuation also requires analysts to distinguish between intermediate and final services. Final services affect welfare directly, whereas intermediate services underpin final services but are not valued directly. For example, biochemical processes in wetlands can reduce the delivery of excess nutrients such as nitrogen to nearby waters (an intermediate service), thereby reducing algae growth and improving water clarity (a final service to swimmers and others who prefer clear water). The economic value of a final service, if correctly measured, should incorporate the value of all intermediate inputs used in its production. Some ecosystem services serve as both final and intermediate services.

A wetland ecosystem. Wetlands provide a number of valuable ecosystem services, including carbon sequestration, water purification, habitats for plants and animals, and a reduction in storm damage and flooding in neighbouring areas.

Additionally, ecosystem services can provide different types of values, some related to the use of the services by humans and others unrelated to human use; these are known to economists as use values and nonuse values, respectively. Use values are related to observable behaviours through which people directly or indirectly use or enjoy ecosystem services. Nonuse values are values unrelated to observed behaviours; these include values generated by the knowledge that something in nature exists, that it can be passed on to future generations, or that it is available to benefit other people (i.e., existence, bequest, and altruistic values).

Role in Policy and Management

There are increasing worldwide efforts to incorporate information on ecosystem services into public and private decisions. In the United States, for example, the President's Committee of Advisors on Science and Technology (PCAST) 1998 report, *Teaming with Life: Investing in Science to Understand and Use America's Living Capital*, highlighted the value and decline of the country's living capital and called for the government to develop sustainable strategies for conservation and management of biodiversity and ecosystems. A 2011 PCAST report, *Sustaining Environmental Capital: Protecting Society and the Economy*, recommended actions to integrate ecosystem services information into federal decisions, including increased use of ecosystem services valuation. Those initiatives accompany multiple agency-specific mandates and initiatives to consider ecosystem services.

Payments for ecosystem services (PES) initiatives, which incentivize the provision of ecosystem services by private suppliers, are emerging worldwide, including national programs in countries such as China and Costa Rica. Many other countries have integrated ecosystem services information into governmental planning processes, including multiple examples in the European Union and Latin America. International initiatives include the World Bank's Wealth Accounting and Valuation of Ecosystem Services (WAVES) program and The Economics of Ecosystems and Biodiversity (TEEB) project. Despite progress in such efforts, many areas of ecosystem services analysis and policy integration remain at a proof-of-concept stage; systematic and formal use of information on ecosystem services to guide public and private decisions is not yet commonplace.

Ecological Goods and Services

Ecological goods are the products of the processes and interactions (described above) of natural systems. For example, plants capturing energy from the sun, combined with water and nutrients from the soil and carbon dioxide from the atmosphere, are able to manufacture food for use by all other organisms. The food is considered a good.

The decomposition of plant material (an ecological service), results in soil production (an ecological good). This ecological good allows us to grow food agriculturally or with human assistance. The processes of seed dispersal and pollination of plants by birds and insects (an ecological service), produce plants (an ecological good). The nutrients we derive from plants we consume are ecological goods resulting from many ecological services.

Ecological goods are not only valuable, they are essential to our survival. Our physical, cultural, social and economic lives are dependent upon these ecological goods that include:

- Clean air
- Fresh water
- Food (and the nutrients (vitamins and minerals) derived from plants and animals)
- Fiber
- Timber
- Other raw building materials
- Genetic resources
- Medicines

Ecological Services

The natural world provides us with the essential services we require for life. These services are called ecological services – the interactions among organisms and their natural environments, including the cycling of water and basic nutrients that humans are able to use and capitalize on. Ecological services include:

- Purification of air and water
- Mitigation of floods and droughts
- Detoxification and decomposition of wastes
- Generation and renewal of soil and natural vegetation
- Pollination of crops and natural vegetation
- Control of the vast majority of potential agricultural pests
- Dispersal of seeds and translocation of nutrients
- Maintenance of biodiversity
- Protection from the sun's harmful ultraviolet rays
- Partial stabilization of climate
- Moderation of temperature extremes and the force of winds and waves
- Support of diverse human culture
- Providing aesthetic beauty and intellectual stimulation that lift the human spirit.

Threats to Ecological Goods and Services

Because of their importance, it is extremely important to reduce the threat of irreversible damage to our ecological systems caused by:

- Land-use change and irreversible conversion of landscapes and their ecological functions.
- Disruption of bio-geochemical cycles i.e. nitrogen, carbon and phosphorus cycles.
- Disruption of the water cycle and ground water recharge.
- Invasion by or introduction of exotic (non-native) organisms.
- Toxins, pollutants and human wastes.
- Changes in chemical composition of the atmosphere and ozone depletion.
- Climate change.

Aquatic Ecosystem

Communities of plants and animals living in water are known as aquatic ecosystems. They are divided into two main groups. Freshwater ecosystems are found in water containing low concentrations of salts, from ponds to estuaries. Marine ecosystems are found in the saltwater of seas and oceans. Most of us are not far away from an aquatic ecosystem of some kind, whether it be in the ocean or a local pond.

The nature of an aquatic ecosystem is shaped, as on land, by the availability of food, oxygen, and the prevailing temperature. Added to this is salinity, which is the salt concentration of the water. Aquatic ecosystems in shallow waters, where there is plenty of sunlight, generally tend to be the most productive. Water pollution, generally coming from human activities, comprises the greatest pressure on aquatic ecosystems. For instance, fish can be killed by acid rain in lakes or lack of oxygen where excess nutrients have been dumped in an estuary.

Aquatic ecosystems are the communities of organisms, their surrounding watery environment, and the relationship between the two. Aquatic, or watery, environments are divided into freshwater and marine. Freshwater has less than one gram per litre of dissolved solids, mainly salts, of which sodium chloride is the most important as far as living organisms are concerned. Salinity is the term given to the concentration of salts in a body of water. Saltwater contains more than one gram per litre of dissolved solids.

Freshwater ecosystems are found in ponds, lakes, reservoirs, rivers, and streams. Marine, or saltwater, ecosystems are found in the seas and oceans. Estuaries, which are places where a river meets the sea, such as San Francisco Bay, are part freshwater and part marine in their makeup. The diversity of an aquatic ecosystem depends upon temperature, availability of light, nutrients, oxygen, and salinity. This makes for very different ecosystems in, for instance, a shallow pond in a temperate climate with plenty of light compared to the deep ocean where conditions are dark and cold.

A wide range of plants, animals, and microbes are found in aquatic ecosystems, from bacteria that tolerate high temperatures in hydrothermal vents at the bottom of the ocean to the blue whale, which is the world's largest animal. The smallest are the phytoplankton and zooplankton, the microscopic animals and plants forming the bottom layer of the aquatic food chain. There are also many aquatic invertebrates including worms, insects, and crustaceans. Among the aquatic vertebrates, amphibians, such as frogs, live on land and water, while fish are purely water-dwelling animals. Many birds, such as kingfishers, gulls, and ducks, live on or near water. Whales are one of the few mammals that can live in the marine environment.

Rivers and streams are lotic, or flowing, freshwater environments. Their organisms tend to be small, with flattened bodies, so they are not swept away. Fallen leaves, insects, and other detritus are important food sources. Lakes and reservoirs are lentic, or layered, systerms with still water. The littoral layer, near the edge of a lake, generally has a large ecosystem with organisms that can use both land and water such as dragonflies, frogs, ducks, and turtles.

Estuaries are regions where the river meets the sea, or ocean. They comprise a semi-enclosed coastal area, open to the marine environment. An estuary has different ecosystems, such as a salt marsh, which is farthest from the sea. Mudflats are tidal and are therefore, sometimes aquatic and sometimes terrestrial. The range of species in mudflats tends to be narrow, but there can be many individuals within this range. The channels are the areas of an estuary that connect to the ocean and are generally rich in fish and crustaceans like crabs. Overall, an estuary is a rich ecosystem because it tends to be shallow and there is a good mixing of nutrients.

The main differences between freshwater and marine environments are the salinity of the water, the depth of the water, and the availability of sunlight for photosynthesis. Freshwater organisms need mechanisms to prevent water loss, because water flows from regions of high salt concentration to those of low salt concentration. Marine organisms at depth have to deal with the pressure of water on their bodies and lack of sunlight. These factors mean that freshwater and marine ecosystems can be very different in their diversity.

Thousands of different invertebrates are found in the seas and oceans. There are sponges, echinoderms (which include starfish, sea urchins, and sea cucumbers), and cnidarians, such as jellyfish, corals, and sea anemones. Mollusks, which are found on land and sea, include the gastropods (limpets, slugs, and snails), cephalopods (octopus, squid, and cuttlefish), and bivalves (clams, mussels, cockles, oysters, and scallops). Most crustaceans are marine animals and include crabs, lobsters, shrimps, barnacles, and woodlice. There are also many species of marine worms living on the sea floor or in the sand of coastal areas.

Around half of the 25,000 known species of fish live in the marine environment, mainly in shallower, warmer.

Marine Ecosystem

Marine ecosystem is a complex of living organisms in the ocean environment.

Marine waters cover two-thirds of the surface of the Earth. In some places the ocean is deeper than Mount Everest is high; for example, the Mariana Trench and the Tonga Trench in the western part of the Pacific Ocean reach depths in excess of 10,000 metres 32,800 feet. Within this ocean habitat live a wide variety of organisms that have evolved in response to various features of their environs.

The Earth formed approximately 4.5 billion years ago. As it cooled, water in the atmosphere condensed and the Earth was pummeled with torrential rains, which filled its great basins, forming seas. The primeval atmosphere and waters harboured the inorganic components hydrogen, methane, ammonia, and water. These substances are thought to have combined to form the first organic compounds when sparked by electrical discharges of lightning. Some of the earliest known organisms are cyanobacteria formerly referred to as blue-green algae. Evidence of these early photosynthetic prokaryotes has been found in Australia in Precambrian marine sediments called stromatolites that are approximately 3 billion years old. Although the diversity of life-forms observed in modern oceans did not appear until much later, during the Precambrian about 4.6 billion to 542 million years ago many kinds of bacteria, algae, protozoa, and primitive metazoa evolved to exploit the early marine habitats of the world. During the Cambrian Period about 542 million to 488 million years ago) a major radiation of life occurred in the oceans. Fossils of familiar organisms such as cnidaria e.g., jellyfish, echinoderms e.g., feather stars, precursors of the fishes e.g., the protochordate *Pikaia* from the Burgess Shale of Canada, and other vertebrates are found in marine sediments of this age. The first fossil fishes are found in sediments from the Ordovician Period about 488 million to 444 million years ago. Changes in the physical conditions of the ocean that are thought to have occurred in the Precambrian—an increase in the concentration of oxygen in seawater and a build-up of the ozone layer that reduced dangerous ultraviolet radiation—may have facilitated the increase and dispersal of living things.

The Marine Environment

Geography, Oceanography, and Topography

The shape of the oceans and seas of the world has changed significantly throughout the past 600 million years. According to the theory of plate tectonics, the crust of the Earth is made up of many dynamic plates. There are two types of plates—oceanic and continental—which float on the surface of the Earth's mantle, diverging, converging, or sliding against one another. When two plates diverge, magma from the mantle wells up and cools, forming new crust; when convergence occurs, one plate descends—i.e., is subducted—below the other and crust is resorbed into the mantle. Examples of both processes are observed in the marine environment. Oceanic crust is created along oceanic ridges or rift areas, which are vast undersea mountain ranges such as the Mid-Atlantic Ridge. Excess crust is reabsorbed along subduction zones, which usually are marked by deep-sea trenches such as the Kuril Trench off the coast of Japan.

The shape of the ocean also is altered as sea levels change. During ice ages a higher proportion of the waters of the Earth is bound in the polar ice caps, resulting in a relatively low sea level. When the polar ice caps melt during interglacial periods, the sea level rises. These changes in sea level cause great changes in the distribution of marine environments such as coral reefs. For example, during the last Pleistocene Ice Age the Great Barrier Reef did not exist as it does today; the continental shelf on which the reef now is found was above the high-tide mark.

Marine organisms are not distributed evenly throughout the oceans. Variations in characteristics of the marine environment create different habitats and influence what types of organisms will inhabit them. The availability of light, water depth, proximity to land, and topographic complexity all affect marine habitats.

The availability of light affects which organisms can inhabit a certain area of a marine ecosystem. The greater the depth of the water, the less light can penetrate until below a certain depth there is no light whatsoever. This area of inky darkness, which occupies the great bulk of the ocean, is called the aphotic zone. The illuminated region above it is called the photic zone, within which are distinguished the euphotic and disphotic zones. The euphotic zone is the layer closer to the surface that receives enough light for photosynthesis to occur. Beneath lies the disphotic zone, which is illuminated but so poorly that rates of respiration exceed those of photosynthesis. The actual depth of these zones depends on local conditions of cloud cover, water turbidity, and ocean surface. In general, the euphotic zone can extend to depths of 80 to 100 metres and the disphotic zone to depths of 80 to 700 metres. Marine organisms are particularly abundant in the photic zone, especially the euphotic portion; however, many organisms inhabit the aphotic zone and migrate vertically to the photic zone every night. Other organisms, such as the tripod fish and some species of sea cucumbers and brittle stars, remain in darkness all their lives.

Marine environments can be characterized broadly as a water, or pelagic, environment and a bottom, or benthic, environment. Within the pelagic environment the waters are divided into the neritic province, which includes the water above the continental shelf, and the oceanic province, which includes all the open waters beyond the continental shelf. The high nutrient levels of the neritic province—resulting from dissolved materials in riverine runoff—distinguish this province from the oceanic. The upper portion of both the neritic and oceanic waters—the epipelagic zone—is where photosynthesis occurs; it is roughly equivalent to the photic zone. Below this zone lie the mesopelagic, ranging between 200 and 1,000 metres, the bathypelagic, from 1,000 to 4,000 metres, and the abyssalpelagic, which encompasses the deepest parts of the oceans from 4,000 metres to the recesses of the deep-sea trenches.

The benthic environment also is divided into different zones. The supralittoral is above the high-tide mark and is usually not under water. The intertidal, or littoral, zone ranges from the high-tide mark (the maximum elevation of the tide) to the shallow, offshore waters. The sublittoral is the environment beyond the low-tide mark and is often used to refer to substrata of the continental shelf, which reaches depths of between 150 and 300 metres. Sediments of the continental shelf that influence marine organisms generally originate from the land, particularly in the form of riverine runoff, and include clay, silt, and sand. Beyond the continental shelf is the bathyal zone, which occurs at depths of 150 to 4,000 metres and includes the descending continental slope and rise. The abyssal zone (between 4,000 and 6,000 metres) represents a substantial portion of the oceans. The deepest region of the oceans (greater than 6,000 metres) is the hadal zone of the deep-sea trenches. Sediments of the deep sea primarily originate from a rain of dead marine organisms and their wastes.

Physical and Chemical Properties of Seawater

The physical and chemical properties of seawater vary according to latitude, depth, nearness to land, and input of fresh water. Approximately 3.5 percent of seawater is composed of dissolved compounds, while the other 96.5 percent is pure water. The chemical composition of seawater reflects such processes as erosion of rock and sediments, volcanic activity, gas exchange with the atmosphere, the metabolic and breakdown products of organisms, and rain. For a list of the principal constituents of seawater. Dissolved inorganic substances. In addition to carbon, the nutrients essential for living organisms include nitrogen and phosphorus, which are minor constituents of seawater and thus are often limiting factors in organic cycles of the ocean. Concentrations of phosphorus and nitrogen are generally low in the photic zone because they are rapidly taken up by marine organisms. The highest concentrations of these nutrients generally are found below 500 metres, a result of the decay of organisms. Other important elements include silicon used in the skeletons of radiolarians and diatoms; and calcium essential in the skeletons of many organisms such as fish and corals).

The chemical composition of the atmosphere also affects that of the ocean. For example, carbon dioxide is absorbed by the ocean and oxygen is released to the atmosphere through the activities of marine plants. The dumping of pollutants into the sea also can affect the chemical makeup of the ocean, contrary to earlier assumptions that, for example, toxins could be safely disposed of there.

The physical and chemical properties of seawater have a great effect on organisms, varying especially with the size of the creature. As an example, seawater is viscous to very small animals less than 1 millimetre [0.039 inch] long such as ciliates but not to large marine creatures such as tuna.

Marine organisms have evolved a wide variety of unique physiological and morphological features that allow them to live in the sea. Notothenid fishes in Antarctica are able to inhabit waters as cold as −2° C (28° F) because of proteins in their blood that act as antifreeze. Many organisms are able to achieve neutral buoyancy by secreting gas into internal chambers, as cephalopods do, or into swim bladders, as some fish do; other organisms use lipids, which are less dense than water, to achieve this effect. Some animals, especially those in the aphotic zone, generate light to attract prey. Animals in the disphotic zone such as hatchet fish produce light by means of organs called photophores to break up the silhouette of their bodies and avoid visual detection by predators. Many marine animals can detect vibrations or sound in the water over great distances by means of specialized organs. Certain fishes have lateral line systems, which they use to detect prey, and whales have a sound-producing organ called a melon with which they communicate. Tolerance to differences in salinity varies greatly: stenohaline organisms have a low tolerance to salinity changes, whereas euryhaline organisms, which are found in areas where river and sea meet estuaries, are very tolerant of large changes in salinity. Euryhaline organisms are also very tolerant of changes in temperature. Animals that migrate between fresh water and salt water, such as salmon or eels, are capable of controlling their osmotic environment by active pumping or the retention of salts. Body architecture varies greatly in marine waters. The body shape of the cnidarian by-the-wind-sailor Velella velella—an animal that lives on the surface of the water pleuston and sails with the assistance of a modified flotation chamber—contrasts sharply with the sleek, elongated shape of the barracuda.

Ocean Currents

The movements of ocean waters are influenced by numerous factors, including the rotation of the Earth which is responsible for the Coriolis effect, atmospheric circulation patterns that influence surface waters, and temperature and salinity gradients between the tropics and the polar regions thermohaline circulation. For a detailed discussion of ocean circulation. The resultant patterns of circulation range from those that cover great areas, such as the North Subtropical Gyre, which follows a path thousands of kilometres long, to small-scale turbulences of less than one metre.

Marine organisms of all sizes are influenced by these patterns, which can determine the range of a species. For example, krill *Euphausia superba* are restricted to the Antarctic Circumpolar Current. Distribution patterns of both large and small pelagic organisms are affected as well. Mainstream currents such as the Gulf Stream and East Australian Current transport larvae great distances. As a result cold temperate coral reefs receive a tropical infusion when fish and invertebrate larvae from the tropics are relocated to high latitudes by these currents. The successful recruitment of eels to Europe depends on the strength of the Gulf Stream to transport them from spawning sites in the Caribbean. Areas where the ocean is affected by near shore features, such as estuaries, or areas in which there is a vertical salinity gradient halocline often exhibit intense biological activity. In these environments, small organisms can become concentrated, providing a rich supply of food for other animals.

Marine Biota

Marine biota can be classified broadly into those organisms living in either the pelagic environment (plankton and nekton) or the benthic environment (benthos). Some organisms, however, are benthic in one stage of life and pelagic in another. Producers that synthesize organic molecules exist in both environments. Single-celled or multicelled plankton with photosyntheticpigments are the producers of the photic zone in the pelagic environment. Typical benthic producers are microalgae (e.g., diatoms), macroalgae (e.g., the kelp *Macrocystis pyrifera*), or sea grass (e.g., *Zostera*).

Plankton

Plankton are the numerous, primarily microscopic inhabitants of the pelagic environment. They are critical components of food chains in all marine environments on community ecology because they provide nutrition for the nekton (e.g., crustaceans, fish, and squid and benthos e.g., sea squirts and sponges. They also exert a global effect on the biosphere because the balance of components of the Earth's atmosphere depends to a great extent on the photosynthetic activities of some plankton.

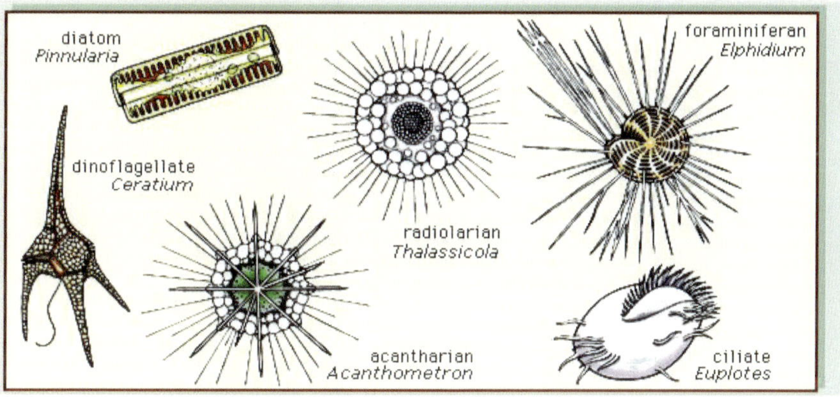

The term plankton is derived from the Greek planktos, meaning wandering or drifting, an apt description of the way most plankton spend their existence, floating with the ocean's currents. Not all plankton, however, are unable to control their movements, and many forms depend on self-directed motions for their survival.

Plankton range in size from tiny microbes 1 micrometre [0.000039 inch] or less to jellyfish whose gelatinous bell can reach up to 2 metres in width and whose tentacles can extend over 15 metres. However, most planktonic organisms, called plankters, are less than 1 millimetre (0.039 inch) long. These microbes thrive on nutrients in seawater and are often photosynthetic. The plankton includes a wide variety of organisms such as algae, bacteria, protozoans, the larvae of some animals, and crustaceans. A large proportion of the plankton are protists—i.e., eukaryotic, predominantly single-celled organisms. Plankton can be broadly divided into phytoplankton, which are plants or plantlike protists; zooplankton, which are animals or animal-like protists; and microbes such as bacteria. Phytoplankton carry out photosynthesis and are the producers of the marine community; zooplankton are the heterotrophic consumers.

Diatoms and dinoflagellates (approximate range between 15 and 1,000 micrometres in length) are two highly diverse groups of photosynthetic protists that are important components of the plankton. Diatoms are the most abundant phytoplankton. While many dinoflagellates carry out photosynthesis, some also consume bacteria or algae. Other important groups of protists include flagellates, foraminiferans, radiolarians, acantharians, and ciliates. Many of these protists are important consumers and a food source for zooplankton.

Zooplankton, which are greater than 0.05 millimetre in size, are divided into two general categories: meroplankton, which spend only a part of their life cycle—usually the larval or juvenile stage—as plankton, and holoplankton, which exist as plankton all their lives. Many larval meroplankton in coastal, oceanic, and even freshwater environments (including sea urchins, intertidal snails, and crabs, lobsters, and fish) bear little or no resemblance to their adult forms. These larvae may exhibit features unique to the larval stage, such as the spectacular spiny armour on the larvae of certain crustaceans e.g., Squilla, probably used to ward off predators.

Important holoplanktonic animals include such lobsterlike crustaceans as the copepods, cladocerans, and euphausids (krill), which are important components of the marine environment because they serve as food sources for fish and marine mammals. Gelatinous forms such as larvaceans, salps, and siphonophores graze on phytoplankton or other zooplankton. Some omnivorous zooplankton such as euphausids and some copepods consume both phytoplankton and zooplankton; their feeding behaviour changes according to the availability and type of prey. The grazing and predatory activity of some zooplankton can be so intense that measurable reductions in phytoplankton or zooplankton abundance (or biomass) occur. For example, when jellyfish occur in high concentration in enclosed seas, they may consume such large numbers of fish larvae as to greatly reduce fish populations.

The jellylike plankton are numerous and predatory. They secure their prey with stinging cells (nematocysts) or sticky cells colloblasts of comb jellies. Large numbers of the Portuguese man-of-war Physalia, with its conspicuous gas bladder, the by-the-wind-sailor Velella velella, and the small blue disk-shaped Porpita porpita are propelled along the surface by the wind, and after strong onshore winds they may be found strewn on the beach. Beneath the surface, comb jellies often abound, as do siphonophores, salps, and scyphomedusae.

The pelagic environment was once thought to present few distinct habitats, in contrast to the array of niches within the benthic environment. Because of its apparent uniformity, the pelagic realm was understood to be distinguished simply by plankton of different sizes. Small-scale variations in the pelagic environment, however, have been discovered that affect biotic distributions. Living and dead matter form organic aggregates called marine snow to which members of the plankton community may adhere, producing patchiness in biotic distributions. Marine snow includes structures such as aggregates of cells and mucus as well as drifting macroalgae and other flotsam that range in size from 0.5 millimetres to 1 centimetre although these aggregates can be as small as 0.05 millimetres and as large as 100 centimetres. Many types of microbes, phytoplankton, and zooplankton stick to marine snow, and some grazing copepods and predators will feed from the surface of these structures. Marine snow is extremely abundant at times, particularly after plankton blooms. Significant quantities of organic material from upper layers of the ocean may sink to the ocean floor as marine snow, providing an important source of food for bottom dwellers. Other structures that plankton respond to in the marine environment include aggregates of phytoplankton cells that form large rafts in tropical and temperate waters of the world e.g., cells of Oscillatoria Trichodesmium erthraeus and various types of seaweed e.g., Sargassum, Phyllospora, Macrocystis that detach from the seafloor and drift.

Nekton

Nekton are the active swimmers of the oceans and are often the best-known organisms of marine waters. Nekton are the top predators in most marine food chains of the community ecology. The distinction between nekton and plankton is not always sharp. As mentioned above, many large marine animals, such as marlin and tuna, spend the larval stage of their lives as plankton and their adult stage as large and active members of the nekton. Other organisms such as krill are referred to as both micronekton and macrozooplankton.

The vast majority of nekton are vertebrates e.g., fishes, reptiles, and mammals, mollusks, and crustaceans. The most numerous group of nekton are the fishes, with approximately 16,000 species. Nekton are found at all depths and latitudes of marine waters. Whales, penguins, seals, and icefish abound in polar waters. Lantern fish family Myctophidae are common in the aphotic zone along with gulpers Saccopharynx, whalefish family Cetomimidae, seven-gilled sharks, and others. Nekton diversity is greatest in tropical waters, where in particular there are large numbers of fish species.

The largest animals on the Earth, the blue whales Balaenoptera musculus, which grow to 25 to 30 metres long, are members of the nekton. These huge mammals and other baleen whales (order Mysticeti), which are distinguished by fine filtering plates in their mouths, feed on plankton and micronekton as do whale sharks Rhinocodon typus, the largest fish in the world (usually 12 to 14 metres long, with some reaching 17 metres). The largest carnivores that consume large prey include the toothed whales (order Odontoceti—for example, the killer whales, Orcinus orca, great white sharks Carcharodon carcharias, tiger sharks Galeocerdo cuvier, black marlin Makaira indica, bluefin tuna Thunnus thynnus, and giant groupers Epinephelus lanceolatus.

Nekton form the basis of important fisheries around the world. Vast schools of small anchovies, herring, and sardines generally account for one-quarter to one-third of the annual harvest from the ocean. Squid are also economically valuable nekton. Halibut, sole, and cod are demersal i.e., bottom-dwelling fish that are commercially important as food for humans. They are generally caught in continental shelf waters. Because pelagic nekton often abound in areas of upwelling where the waters are nutrient-rich, these regions also are major fishing areas.

Benthos

Organisms are abundant in surface sediments of the continental shelf and in deeper waters, with a great diversity found in or on sediments. In shallow waters, beds of seagrass provide a rich habitat for polychaete worms, crustaceans e.g., amphipods, and fishes. On the surface of and within intertidal sediments most animal activities are influenced strongly by the state of the tide. On many sediments in the photic zone, however, the only photosynthetic organisms are microscopic benthic diatoms.

Benthic organisms can be classified according to size. The macrobenthos are those organisms larger than 1 millimetre. Those that eat organic material in sediments are called deposit feeders e.g., holothurians, echinoids, gastropods, those that feed on the plankton above are the suspension feeders e.g., bivalves, ophiuroids, crinoids, and those that consume other fauna in the benthic assemblage are predators (e.g., starfish, gastropods). Organisms between 0.1 and 1 millimetre constitute the meiobenthos. These larger microbes, which include foraminiferans, turbellarians, and polychaetes, frequently dominate benthic food chains, filling the roles of nutrient recycler, decomposer, primary producer, and predator. The microbenthos are those organisms smaller than 1 millimetre; they include diatoms, bacteria, and ciliates.

Organic matter is decomposed aerobically by bacteria near the surface of the sediment where oxygen is abundant. The consumption of oxygen at this level, however, deprives deeper layers of oxygen, and marine sediments below the surface layer are anaerobic. The thickness of the oxygenated layer varies according to grain size, which determines how permeable the sediment is to oxygen and the amount of organic matter it contains. As oxygen concentration diminishes, anaerobic processes come to dominate. The transition layer between oxygen-rich and oxygen-poor layers is called the redox discontinuity layer and appears as a gray layer above the black anaerobic layers. Organisms have evolved various ways of coping with the lack of oxygen. Some anaerobes release hydrogen sulfide, ammonia, and other toxic reduced ions through metabolic processes. The thiobiota, made up primarily of microorganisms, metabolize sulfur. Most organisms that live below the redox layer, however, have to create an aerobic environment for themselves. Burrowing animals generate a respiratory current along their burrow systems to oxygenate their dwelling places; the influx of oxygen must be constantly maintained because the surrounding anoxic layer quickly depletes the burrow of oxygen. Many bivalves e.g., Mya arenaria extend long siphons upward into oxygenated waters near the surface so that they can respire and feed while remaining sheltered from predation deep in the sediment. Many large mollusks use a muscular "foot" to dig with, and in some cases they use it to propel themselves away from predators such as starfish. The consequent "irrigation" of burrow systems can create oxygen and nutrient fluxes that stimulate the production of benthic producers e.g., diatoms.

Not all benthic organisms live within the sediment; certain benthic assemblages live on a rocky substrate. Various phyla of algae—Rhodophyta red, Chlorophyta green, and Phaeophyta brown—are abundant and diverse in the photic zone on rocky substrata and are important producers. In intertidal regions algae are most abundant and largest near the low-tide mark. Ephemeral algae such as Ulva, Enteromorpha, and coralline algae cover a broad range of the intertidal. The mix of algae species found in any particular locale is dependent on latitude and also varies greatly according to wave exposure and the activity of grazers. For example, Ascophyllum spores cannot attach to rock in even a gentle ocean surge; as a result this plant is largely restricted to sheltered shores. The fastest-growing plant—adding as much as 1 metre per day to its length—is the giant kelp, Macrocystis pyrifera, which is found on subtidal rocky reefs. These plants, which may exceed 30 metres in length, characterize benthic habitats on many temperate reefs. Large laminarian and fucoid algae are also common on temperate rocky reefs, along with the encrusting (e.g., Lithothamnion) or short tufting forms e.g., Pterocladia. Many algae on rocky reefs are harvested for food, fertilizer, and pharmaceuticals. Macroalgae are relatively rare on tropical reefs where corals abound, but Sargassum and a diverse assemblage of short filamentous and tufting algae are found, especially at the reef crest. Sessile and slow-moving invertebrates are common on reefs. In the intertidal and subtidal regions herbivorous gastropods and urchins abound and can have a great influence on the distribution of algae. Barnacles are common sessile animals in the intertidal. In the subtidal regions, sponges, ascidians, urchins, and anemones are particularly common where light levels drop and current speeds are high. Sessile assemblages of animals are often rich and diverse in caves and under boulders.

Reef-building coral polyps Scleractinia are organisms of the phylum Cnidaria that create a calcareous substrate upon which a diverse array of organisms live. Approximately 700 species of corals are found in the Pacific and Indian oceans and belong to genera such as Porites, Acropora, and Montipora. Some of the world's most complex ecosystems are found on coral reefs. Zooxanthellae are the photosynthetic, single-celled algae that live symbiotically within the tissue of corals and help to build the solid calcium carbonate matrix of the reef. Reef-building corals are found only in waters warmer than 18° C; warm temperatures are necessary, along with high light intensity, for the coral-algae complex to secrete calcium carbonate. Many tropical islands are composed entirely of hundreds of metres of coral built atop volcanic rock.

Links between the Pelagic Environments and the Benthos

Considering the pelagic and benthic environments in isolation from each other should be done cautiously because the two are interlinked in many ways. For example, pelagic plankton are an important source of food for animals on soft or rocky bottoms. Suspension feeders such as anemones and barnacles filter living and dead particles from the surrounding water while detritus feeders graze on the accumulation of particulate material raining from the water column above. The molts of crustaceans, plankton feces, dead plankton, and marine snow all contribute to this rain of fallout from the pelagic environment to the ocean bottom. This fallout can be so intense in certain weather patterns—such as the El Niño condition—that benthic animals on soft bottoms are smothered and die. There also is variation in the rate of fallout of the plankton according to seasonal cycles of production. This variation can create seasonality in the abiotic zone where there is little or no variation in temperature or light. Plankton form marine sediments, and many types of fossilized protistan plankton, such as foraminiferans and coccoliths, are used to determine the age and origin of rocks.

Organisms of the Deep Sea Vents

hydrothermal mussels: Galatheid crabs and shrimp grazing on the bacterial filaments that grow on the shells of the hydrothermal mussels covering the Northwest Eifuku volcano in the Mariana Arc region.

Producers were discovered in the aphotic zone when exploration of the deep sea by submarine became common in the 1970s. Deep-sea hydrothermal vents now are known to be relatively common in areas of tectonic activity e.g., spreading ridges. The vents are a nonphotosynthetic source of organic carbon available to organisms. A diversity of deep-sea organisms including mussels, large bivalve clams, and vestimentiferan worms are supported by bacteria that oxidize sulfur sulfide and derive chemical energy from the reaction. These organisms are referred to as chemoautotrophic, or chemosynthetic, as opposed to photosynthetic, organisms. Many of the species in the vent fauna have developed symbiotic relationships with chemoautotrophic bacteria, and as a consequence the megafauna are principally responsible for the primary production in the vent assemblage. The situation is analogous to that found on coral reefs where individual coral polyps have symbiotic relationships with zooxanthellae. In addition to symbiotic bacteria there is a rich assemblage of free-living bacteria around vents. For example, Beggiatoas-like bacteria often form conspicuous weblike mats on any hard surface; these mats have been shown to have chemoautotrophic metabolism. Large numbers of brachyuran e.g., Bythograea and galatheid crabs, large sea anemones e.g., Actinostola callasi, copepods, other plankton, and some fish—especially the eelpout Thermarces cerberus—are found in association with vents.

Freshwater Ecosystem

Fresh water is best defined, in contrast to the oceans, as water that contains a relatively small amount of dissolved chemical compounds. Some studies of fresh-water ecosystems focus on water bodies themselves, while others include the surrounding land that interacts with a lake or stream

Fresh-water ecosystems are often categorized by two basic criteria: water movement and size. In lotic or flowing-waterecosystems the water moves steadily in a uniform direction, while in lentic or standing-water systems the water tends toremain in the same general area for a longer period of time. Size varies dramatically in each category. Lotic systems rangefrom a tiny rivulet dripping off a rock to large rivers. Lentic systems range from the water borne within a cup formed by smallplants or tree holes to very large water bodies such as the Laurentian Great Lakes. Fresh-water studies also consider theinteractions of the geological, physical, and chemical features along with the biota, the organisms that occur in an area.

Physical Environment

The quantity and spectral quality of light have major influences on the distribution of the biota and also play a central role inthe thermal structure of lakes. The light that reaches the surface of a lake or stream is controlled by latitude, season, time ofday, weather, and the conditions that surround a water body. Light penetration is controlled by the nature of water itself andby dissolved and particulate material in a water column.

Water exhibits a number of unusual thermal properties, including its existence in liquid state at normal earth surfacetemperatures, a remarkable ability to absorb heat, and a maximum density at 39.09 °F (3.94 °C)' which leads to a complexannual cycle in the temperature structure of fresh-water ecosystems.

As water is warmed at the surface of a lake, a stable condition is reached in which a physically distinct upper layer of water,the epilimnion, is maintained over a deeper, cooler stratum, the hypolimnion. The region of sharp temperature changesbetween these two layers is called the metalimnion. The characteristic establishment of two layers is of major importance inthe chemical cycling within lakes and consequently for the biota.

As the surface waters of a lake cool, the density of epilimnetic waters increases, which decreases their resistance to mixingwith the hypolimnion. If cooling continues, the entire water column will mix, an event known as turnover. At temperaturesbelow 39.09 °F (3.94 °C)' water again becomes less dense; ice and very cold water float above slightly warmer water, maintaining liquid water below ice cover even in lakes in the Antarctic. Many lakes in the temperate zones undergo twodistinct periods of mixing annually, one in the spring and the other in the fall, that separate periods of stratification in thesummer and winter.

Water movement is more extensive in lotic than in standing-water ecosystems, but water motion has important effects in bothtypes. Turbulence occurs ubiquitously and affects the distribution of organisms, particles, dissolved substances, and heat.Turbulence increases with the velocity of flowing water, and the amount of material transported by water increases withturbulence. Flowing-water ecosystems are characterized by large fluctuations in the velocity and amount of water. Aside fromsurface waves on large lakes, most water movement in lentic systems is not conspicuous.

Chemical Environment

For an element, three basic parameters are of importance: the forms in which it occurs, its source, and its concentration in water relative to its biological demand or effect. Most elements are derived from dissolved gases in the atmosphere or from minerals in geological materials surrounding a lake. In some cases the presence of elements is strongly mediated by biological activities.

Oxygen occurs as dissolved O_2 and in combination with other elements resulting from chemical or biological reactions. It enters water primarily from the atmosphere through a combination of diffusion and turbulent mixing. When biological demands for oxygen exceed supply rates, it can be depleted from fresh-water ecosystems. Anoxic conditions occur in hypolimnia during summer and under ice cover in winter when lake strata are isolated from the atmosphere. Oxygen depletion may also occur in rivers that receive heavy organic loading. Aside from specialized bacteria, few organisms can occur under anoxic conditions.

Carbon dioxide is derived primarily from the atmosphere, with additional sources from plant and animal respiration and carbonate minerals. Its chemical species exert a major control on the hydrogen ion concentration of water (the acidity or pH).

Phosphorus occurs primarily as a phosphate ion or in a number of complex organic forms. It is the element which is most commonly in the shortest supply relative to biological demand. Phosphorus is thus a limiting nutrient, and its addition to fresh-water ecosystems through human activities can lead to major problems due to increased growth of aquatic plants.

Nitrogen occurs in water as N_2, NO_2, NO_3, NH_4, and in diverse organic forms. It may be derived from precipitation and soils, but its availability is usually regulated by bacterial processes. Nitrogen occurs in relatively short supply relative to biological demand. It may also limit growth in some fresh-water systems, particularly when phosphorus levels have been increased because of human activity.

A variety of other elements also help determine the occurrence of fresh-water organisms either directly or by the elements' effects on water chemistry.

Biota

In addition to taxonomy, fresh-water organisms are classified by the areas in which they occur, the manner in which theymove, and the roles that they occupy in trophic webs. Major distinctions are made between organisms that occur in bottomareas and those within the water column, the limnetic zone. Production is the most difficult variable to measure, but itprovides the greatest information on the role of organisms in an ecosystem.

Plankton organisms occur in open water and move primarily with general water motion. Planktonic communities occur in all lentic ecosystems. In lotic systems they are important only in slow-moving areas.

Phytoplankton (plant plankton) comprise at least eight major taxonomic groups of algae, most of which are microscopic. They exhibit a diversity of forms ranging from one-celled organisms to complex colonies.

Zooplankton (animal plankton) comprise protozoans and three major groups of eukaryotic organisms: rotifers, cladocerans, and copepods. Most are microscopic but some are clearly visible to the naked eye.

Animals, such as fishes and swimming insects, that occur in the water column and can control their position independently of water movement are termed nekton. In addition to their importance as a human food source, fishes may affect zooplankton, benthic invertebrates, vegetation, and lake sediments.

Benthic organisms are a diverse group associated with the bottoms of lakes and streams. The phytobenthos ranges from microscopic algae to higher plants. Benthic animals range from microscopic protozoans and crustaceans to large aquatic insects and fishes

Bacteria occur throughout fresh-water ecosystems in planktonic and benthic areas and play a major role in biogeochemical cycling. Most bacteria are heterotrophic, using reduced carbon as an energy source; others are photosynthetic or derive energy from reduced compounds other than carbon.

Interactions

Ultimately the conditions in a fresh-water ecosystem are controlled by numerous interactions among biotic and abioticcomponents. Primary production in a fresh-water ecosystem is controlled by light and nutrient availability. As light diminisheswith depth in a column or water, a point is reached where energy for photosynthesis balances respiratory energy demands.In benthic areas, the region where light is sufficient for plant growth is termed the littoral zone; deeper areas are labeled profundal.

Nutrient availability generally controls the total amount of primary production that occurs in fresh-water ecosystems. One classification scheme for lakes ranks them according to total production, ranging from oligotrophic lakes, where water is clearand production is low, to eutrophic systems, characterized by high nutrient concentrations, high standing algal biomass, highproduction, low water clarity, and low concentrations of oxygen in the hypolimnion. Eutrophic conditions are more likely tooccur as a lake ages. This aging process, termed eutrophication, occurs naturally but can be greatly accelerated byanthropogenic additions of nutrients. A third major lake category, termed dystrophy, occurs when large amounts of organicmaterials that are resistant to decomposition wash into a lake basin. These organic materials stain the lake water and have amajor influence on water chemistry which results in low production.

Lentic

Lentic water systems consist of still bodies of water, such as lakes, ponds, and seas. During periods of drought these systems will often last longer than their smaller counterparts and organisms can continue to live despite the shortened supplies. These bodies, often deep with multiple zones experience many things that lotic water systems do not. Lentic water systems are made up of multiple zones, littoral, limnetic, vertical, and benthic.

The epilimnion, or surface water, is the area in which the most life can be found. This zone is high in oxygen content during the summers, the warmer waters that receive the most sunlight will hold the most oxygen, thus allowing the greatest amount of life to flourish here.

The metalimnion, or middle mass of water is where the temperature of the water begins to decrease, life is not as plentiful as in the surface waters but there are still many organisms that can be found here. Often times this is where the thermocline will begin, a thermocline is a point at which the water temperature decreases approximately one degree Celsius per meter.

The hypolimnion is the warmest zone during the winter time and the coldest during the summer. It is the bottom most part of the body, where sometimes light does not reach. This zone is where the least amount of life will be found throughout most of the year.

Overturns occur when water is stratified in bodies, the water from the bottom of the mass is mixed with the water close to the surface. This occurs during spring and fall.

A. Inland freshwater areas	
1. Seasonally flooded	Soil covered with water or waterlogged during variable periods but well-drained during much of the growing season. In upland depressions and bottomlands. Bottomland hardwoods to herbaceous growth.
2. Freshwater meadows	Without standing water during growing season; water logged to within few inches of surface. Grasses, rushes, sedges, broad leaf plants.
3. Shallow fresh water marshes	Soil waterlogged during growing season; often covered with 6 inches or more of water. Grasses, bulrushes, spike rushes, cattails, arrowhead, smartweed, pickerelweed. A major water-fowl-production area.
4. Deep freshwater marshes	Soil covered with 6 inches to 3 feet of water Cattails, reeds, bulrushes, spikes rushes, wild rice. Principal duck-breeding area.

5. Open freshwater marshes	Water less than 10 feet deep. Bordered by emergent vegetation: pondweed, naiads, wild celery, and water lily. Brooding, feeding, and nesting areas of ducks.
6. Shrub swamps	Soil waterlogged; often covered with 6 inches or more of water. Alder, willow, buttonbush, dogwoods. Ducks nesting and feeding to limited extent.
7. Wooded swamps	Soil waterlogged; often covered with 1 foot of water. Along sluggish streams, flat uplands shallow lakes, basins. North of USA: tamarack, arborvitae, spruce, red maple, silver maple. South of USA; water oak, overcup oak, tupelo, swamp black gum cypress.
8 Bogs	Soil waterlogged; spongy covering of mosses. Heath shrubs, sphagnum, sedges.
B. Coastal freshwater areas	
9. Shallow freshwater marsh	Soil waterlogged during growing season; at high tides as much as 6 inches of water on landward side, deep marshes along tidal rivers, sounds, deltas. Grasses and sedges. Important water-fowl areas.
10. Deep freshwater marshes	At high tide covered with 6 inches to 6 feet of water. Along tidal rivers and bays. Cattails, wild rice, giant cutgrass.
11 Open freshwater	Shallow portions of open water along fresh tidal rivers and sounds Vegetation scarce or absent. Important water-fowl areas.

Table: Classification of wetland or lentic ecosystems (excluding lakes and ponds)

Important Abiotic Factors

Light

Light provides the solar energy required to drive the process of photosynthesis, the major energy source of lentic systems. The amount of light received depends upon a combination of several factors. Small ponds may experience shading by surrounding trees, while cloud cover may affect light availability in all systems, regardless of size. Seasonal and diurnal considerations also play a role in light availability because the shallower the angle at which light strikes water, the more light is lost by reflection. This is known as Beer's law. Once light has penetrated the surface, it may also be scattered by particles suspended in the water column. This scattering decreases the total amount of light as depth increases. Lakes are divided into photic and aphotic regions, the prior receiving sunlight and latter being below the depths of light penetration, making it void of photosynthetic capacity. In relation to Lake Zonation, the pelagic and benthic zones are considered to lie within the photic region, while the profundal zone is in the aphotic region.

Temperature

Temperature is an important abiotic factor in lentic ecosystems because most of the biota are poikilothermic, where internal body temperatures are defined by the surrounding system. Water can be heated or cooled through radiation at the surface and conduction to or from the air and surrounding substrate. Shallow ponds often have a continuous temperature gradient from warmer waters at the surface to cooler waters at the bottom. In addition, temperature fluctuations can be very great in these systems, both diurnally and seasonally.

Temperature regimes are very different in large lakes. In temperate regions, for example, as air temperatures increase, the icy layer formed on the surface of the lake breaks up, leaving the water at approximately 4 °C. This is the temperature at which water has the highest density. As the season progresses, the warmer air temperatures heat the surface waters, making them less dense. The deeper waters remain cool and dense due to reduced light penetration. As the summer begins, two distinct layers become established, with such a large temperature difference between them that they remain stratified. The lowest zone in the lake is the coldest and is called the hypolimnion. The upper warm zone is called the epilimnion. Between these zones is a band of rapid temperature change called the thermocline. During the colder fall season, heat is lost at the surface and the epilimnion cools. When the temperatures of the two zones are close enough, the waters begin to mix again to create a uniform temperature, an event termed lake turnover. In the winter, inverse stratification occurs as water near the surface cools freezes, while warmer, but denser water remains near the bottom. A thermocline is established, and the cycle repeats.

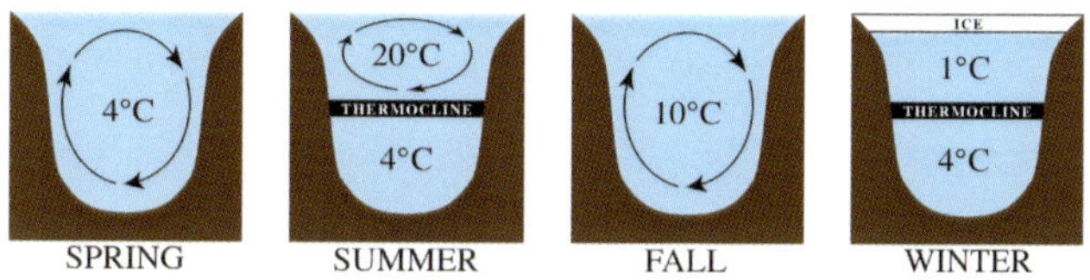

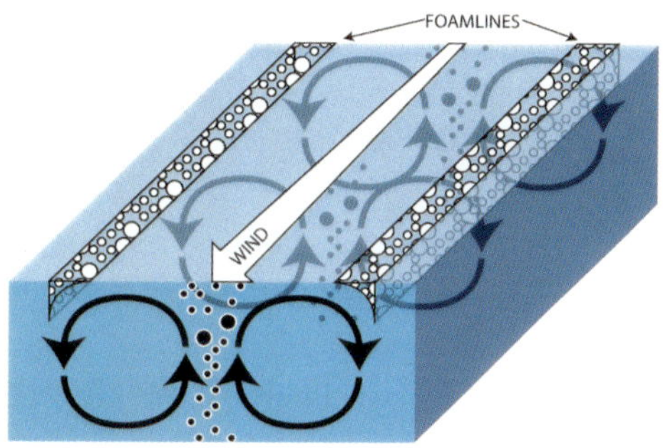

Wind

In exposed systems, wind can create turbulent, spiral-formed surface currents called Langmuir circulations. Exactly how these currents become established is still not well understood, but it is evident that it involves some interaction between horizontal surface currents and surface gravity waves. The visible results of these rotations, which can be seen in any lake, are the surface foam lines that run parallel to the wind direction. Positively buoyant particles and small organisms concentrate in the foam line at the surface and negatively buoyant objects are found in the upwelling current between the two rotations. Objects with neutral buoyancy tend to be evenly distributed in the water column. This turbulence circulates nutrients in the water column, making it crucial for many pelagic species, however its effect on benthic and profundal organisms is minimal to non-existent, respectively. The degree of nutrient circulation is system specific, as it depends upon such factors as wind strength and duration, as well as lake or pool depth and productivity.

Chemistry

Oxygen is essential for organismal respiration. The amount of oxygen present in standing waters depends upon:

1) The area of transparent water exposed to the air,
2) The circulation of water within the system and,

3) The amount of oxygen generated and used by organisms present. In shallow, plant-rich pools there may be great fluctuations of oxygen, with extremely high concentrations occurring during the day due to photosynthesis and very low values at night when respiration is the dominant process of primary producers. Thermal stratification in larger systems can also affect the amount of oxygen present in different zones. The epilimnion is oxygen rich because it circulates quickly, gaining oxygen via contact with the air. The hypolimnion, however, circulates very slowly and has no atmospheric contact. Additionally, fewer green plants exist in the hypolimnion, so there is less oxygen released from photosynthesis. In spring and fall when the epilimnion and hypolimnion mix, oxygen becomes more evenly distributed in the system. Low oxygen levels are characteristic of the profundal zone due to the accumulation of decaying vegetation and animal matter that "rains" down from the pelagic and benthic zones and the inability to support primary producers.

Phosphorus is important for all organisms because it is a component of DNA and RNA and is involved in cell metabolism as a component of ATP and ADP. Also, phosphorus is not found in large quantities in freshwater systems, limiting photosynthesis in primary producers, making it the main determinant of lentic system production. The phosphorus cycle is complex, but the model outlined below describes the basic pathways. Phosphorus mainly enters a pond or lake through runoff from the watershed or by atmospheric deposition. Upon entering the system, a reactive form of phosphorus is usually taken up by algae and macrophytes, which release a non-reactive phosphorus compound as a byproduct of photosynthesis. This phosphorus can drift downwards and become part of the benthic or profundal sediment, or it can be remineralized to the reactive form by microbes in the water column. Similarly, non-reactive phosphorus in the sediment can be remineralized into the reactive form. Sediments are generally richer in phosphorus than lake water, however, indicating that this nutrient may have a long residency time there before it is remineralized and re-introduced to the system.

Lentic System Biota

Bacteria

Bacteria are present in all regions of lentic waters. Free-living forms are associated with decomposing organic material, biofilm on the surfaces of rocks and plants, suspended in the water column, and in the sediments of the benthic and profundal zones. Other forms are also associated with the guts of lentic animals as parasites or in commensal relationships. Bacteria play an important role in system metabolism through nutrient recycling, which is discussed in the Trophic Relationships section.

Primary Producers

Algae, including both phytoplankton and periphyton are the principle photosynthesizers in ponds and lakes. Phytoplankton are found drifting in the water column of the pelagic zone. Many species have a higher density than water which should make them sink and end up in the benthos. To combat this, phytoplankton have developed density changing mechanisms, by forming vacuoles and gas vesicles or by changing their shapes to induce drag, slowing their descent. A very sophisticated adaptation utilized by a small number of species is a tail-like flagellum that can adjust vertical position and allow movement in any direction. Phytoplankton can also maintain their presence in the water column by being circulated in Langmuir rotations. Periphytic algae, on the other hand, are attached to a substrate. In lakes and ponds, they can cover all benthic surfaces. Both types of plankton are important as food sources and as oxygen providers.

Aquatic plants live in both the benthic and pelagic zones and can be grouped according to their manner of growth:

1) Emergent = rooted in the substrate but with leaves and flowers extending into the air,

2) Floating-leaved = rooted in the substrate but with floating leaves,

3) Submersed = growing beneath the surface and,

4) free-floating macrophytes = not rooted in the substrate and floating on the surface. These various forms of macrophytes generally occur in different areas of the benthic zone, with emergent vegetation nearest the shoreline, then floating-leaved macrophytes, followed by submersed vegetation. Free-floating macrophytes can occur anywhere on the system's surface.

Aquatic plants are more buoyant than their terrestrial counterparts because freshwater has a higher density than air. This makes structural rigidity unimportant in lakes and ponds (except in the aerial stems and leaves). Thus, the leaves and stems of most aquatic plants use less energy to construct and maintain woody tissue, investing that energy into fast growth instead. In order to contend with stresses induced by wind and waves, plants must be both flexible and tough. Light, water depth and substrate types are the most important factors controlling the distribution of submerged aquatic plants.Macrophytes are sources of food, oxygen, and habitat structure in the benthic zone, but cannot penetrate the depths of the euphotic zone and hence are not found there.

Invertebrates

Zooplankton are tiny animals suspended in the water column. Like phytoplankton, these species have developed mechanisms that keep them from sinking to deeper waters, including drag inducing body forms and the active flicking of appendages such as antennae or spines. Remaining in the water column may have its advantages in terms of feeding, but this zone's lack of refugia leaves zooplankton vulnerable to predation. In response, some species, especially Daphnia sp., make daily vertical migrations in the water column by passively sinking to the darker lower depths during the day and actively moving towards the surface during the night. Also, because conditions in a lentic system can be quite variable across seasons, zooplankton have the ability to switch from laying regular eggs to resting eggs when there is a lack of food, temperatures fall below 2 °C, or if predator abundance is high. These resting eggs have a diapause, or dormancy period that should allow the zooplankton to encounter conditions that are more favorable to survival when they finally hatch. The invertebrates that inhabit the benthic zone are numerically dominated by small species and are species rich compared to the zooplankton of the open water. They include Crustaceans (e.g. crabs, crayfish, and shrimp), molluscs (e.g. clams and snails), and numerous types of insects. These organisms are mostly found in the areas of macrophyte growth, where the richest resources, highly oxygenated water, and warmest portion of the ecosystem are found. The structurally diverse macrophyte beds are important sites for the accumulation of organic matter, and provide an ideal area for colonization. The sediments and plants also offer a great deal of protection from predatory fishes.

Very few invertebrates are able to inhabit the cold, dark, and oxygen poor profundal zone. Those that can are often red in color due to the presence of large amounts of hemoglobin, which greatly increases the amount of oxygen carried to cells. Because the concentration of oxygen within this zone is low, most species construct tunnels or borrows in which they can hide and make the minimum movements necessary to circulate water through, drawing oxygen to them without expending much energy.

Fish and Other Vertebrates

Fish have a range of physiological tolerances that are dependent upon which species they belong to. They have different lethal temperatures, dissolved oxygen requirements, and spawning needs that are based on their activity levels and behaviors. Because fish are highly mobile, they are able to deal with unsuitable abiotic factors in one zone by simply moving to another. A detrital feeder in the profundal zone, for example, that finds the oxygen concentration has dropped too low may feed closer to the benthic zone. A fish might also alter its residence during different parts of its life history: hatching in a sediment nest, then moving to the weedy benthic zone to develop in a protected environment with food resources, and finally into the pelagic zone as an adult.

Other vertebrate taxa inhabit lentic systems as well. These include amphibians (e.g. salamanders and frogs), reptiles (e.g. snakes, turtles, and alligators), and a large number of waterfowl species. Most of these vertebrates spend part of their time in terrestrial habitats and thus are not directly affected by abiotic factors in the lake or pond. Many fish species are important as consumers and as prey species to the larger vertebrates mentioned above.

Trophic Relationships

Primary Producers

Lentic systems gain most of their energy from photosynthesis performed by aquatic plants and algae. This autochthonous process involves the combination of carbon dioxide, water, and solar energy to produce carbohydrates and dissolved oxygen. Within a lake or pond, the potential rate of photosynthesis generally decreases with depth due to light attenuation. Photosynthesis, however, is often low at the top few millimeters of the surface, likely due to inhibition by ultraviolet light. The exact depth and photosynthetic rate measurements of this curve are system specific and depend upon:

1) The total biomass of photosynthesizing cells,

2) The amount of light attenuating materials and,

3) The abundance and frequency range of light absorbing pigments (i.e. chlorophylls) inside of photosynthesizing cells. The energy created by these primary producers is important for the community because it is transferred to higher trophic levels via consumption.

Bacteria

The vast majority of bacteria in lakes and ponds obtain their energy by decomposing vegetation and animal matter. In the pelagic zone, dead fish and the occasional allochthonous input of litter fall are examples of coarse particulate organic matter (CPOM>1 mm). Bacteria degrade these into fine particulate organic matter (FPOM<1 mm) and then further into usable nutrients. Small organisms such as plankton are also characterized as FPOM. Very low concentrations of nutrients are released during decomposition because the bacteria are utilizing them to build their own biomass. Bacteria, however, are consumed by protozoa, which are in turn consumed by zooplankton, and then further up the trophic levels. Nutrients, including those that contain carbon and phosphorus, are reintroduced into the water column at any number of points along this food chain via excretion or organism death, making them available again for bacteria. This regeneration cycle is known as the microbial loop and is a key component of lentic food webs.

The decomposition of organic materials can continue in the benthic and profundal zones if the matter falls through the water column before being completely digested by the pelagic bacteria. Bacteria are found in the greatest abundance here in sediments, where they are typically 2-1000 times more prevalent than in the water column.

Benthic Invertebrates

Benthic invertebrates, due to their high level of species richness, have many methods of prey capture. Filter feeders create currents via siphons or beating cilia, to pull water and its nutritional contents, towards themselves for straining. Grazers use scraping, rasping, and shredding adaptations to feed on periphytic algae and macrophytes. Members of the collector guild browse the sediments, picking out specific particles with raptorial appendages. Deposit feeding invertebrates indiscriminately consume sediment, digesting any organic material it contains. Finally, some invertebrates belong to the predator guild, capturing and consuming living animals. The profundal zone is home to a unique group of filter feeders that use small body movements to draw a current through burrows that they have created in the sediment. This mode of feeding requires the least amount of motion, allowing these species to conserve energy. A small number of invertebrate taxa are predators in the profundal zone. These species are likely from other regions and only come to these depths to feed. The vast majority of invertebrates in this zone are deposit feeders, getting their energy from the surrounding sediments.

Fish

Fish size, mobility, and sensory capabilities allow them to exploit a broad prey base, covering multiple zonation regions. Like invertebrates, fish feeding habits can be categorized into guilds. In the pelagic zone, herbivores graze on periphyton and macrophytes or pick phytoplankton out of the water column. Carnivores include fishes that feed on zooplankton in the water column (zooplanktivores), insects at the water's surface, on benthic structures, or in the sediment (insectivores), and those that feed on other fish (piscivores). Fish that consume detritus and gain energy by processing its organic material are called detritivores. Omnivores ingest a wide variety of prey, encompassing floral, faunal, and detrital material. Finally, members of the parasitic guild acquire nutrition from a host species, usually another fish or large vertebrate. Fish taxa are flexible in their feeding roles, varying their diets with environmental conditions and prey availability. Many species also undergo a diet shift as they develop. Therefore, it is likely that any single fish occupies multiple feeding guilds within its lifetime.

Lentic Food Webs

As noted in the previous sections, the lentic biota are linked in complex web of trophic relationships. These organisms can be considered to loosely be associated with specific trophic groups (e.g. primary producers, herbivores, primary carnivores, secondary carnivores, etc.). Scientists have developed several theories in order to understand the mechanisms that control the abundance and diversity within these groups. Very generally, top-down processes dictate that the abundance of prey taxa is dependent upon the actions of consumers from higher trophic levels. Typically, these processes operate only between two trophic levels, with no effect on the others. In some cases, however, aquatic systems experience a trophic cascade; for example, this might occur if primary producers experience less grazing by herbivores because these herbivores are suppressed by carnivores. Bottom-up processes are functioning when the abundance or diversity of members of higher trophic levels is dependent upon the availability or quality of resources from lower levels. Finally, a combined regulating theory, bottom-up: top-down, combines the predicted influences of consumers and resource availability. It predicts that trophic levels close to the lowest trophic levels will be most influenced by bottom-up forces, while top-down effects should be strongest at top levels.

Community Patterns and Diversity

Local Species Richness

The biodiversity of a lentic system increases with the surface area of the lake or pond. This is attributable to the higher likelihood of partly terrestrial species of finding a larger system. Also, because larger systems typically have larger populations, the chance of extinction is decreased.

Additional factors, including temperature regime, pH, nutrient availability, habitat complexity, speciation rates, competition, and predation, have been linked to the number of species present within systems.

Succession Patterns in Plankton Communities – The PEG Model

Phytoplankton and zooplankton communities in lake systems undergo seasonal succession in relation to nutrient availability, predation, and competition. Sommer *et al.* described these patterns as part of the Plankton Ecology Group (PEG) model, with 24 statements constructed from the analysis of numerous systems. The following includes a subset of these statements, as explained by Brönmark and Hansson illustrating succession through a single seasonal cycle:

Winter

1. Increased nutrient and light availability result in rapid phytoplankton growth towards the end of winter. The dominant species, such as diatoms, are small and have quick growth capabilities.
2. These plankton are consumed by zooplankton, which become the dominant plankton taxa.

Spring

3. A clear water phase occurs, as phytoplankton populations become depleted due to increased predation by growing numbers of zooplankton.

Summer

4. Zooplankton abundance declines as a result of decreased phytoplankton prey and increased predation by juvenile fishes.
5. With increased nutrient availability and decreased predation from zooplankton, a diverse phytoplankton community develops.
6. As the summer continues, nutrients become depleted in a predictable order: phosphorus, silica, and then nitrogen. The abundance of various phytoplankton species varies in relation to their biological need for these nutrients.
7. Small-sized zooplankton become the dominant type of zooplankton because they are less vulnerable to fish predation.

Fall

8. Predation by fishes is reduced due to lower temperatures and zooplankton of all sizes increase in number.

Winter

9. Cold temperatures and decreased light availability result in lower rates of primary production and decreased phytoplankton populations. 10. Reproduction in zooplankton decreases due to lower temperatures and less prey.

The PEG model presents an idealized version of this succession pattern, while natural systems are known for their variation.

Latitudinal Patterns

There is a well-documented global pattern that correlates decreasing plant and animal diversity with increasing latitude, that is to say, there are fewer species as one moves towards the poles. The cause of this pattern is one of the greatest puzzles for ecologists today. Theories for its explanation include energy availability, climatic variability, disturbance, competition, etc. Despite this global diversity gradient, this pattern can be weak for freshwater systems compared to global marine and terrestrial systems. This may be related to size, as Hillebrand and Azovsky found that smaller organisms (protozoa and plankton) did not follow the expected trend strongly, while larger species (vertebrates) did. They attributed this to better dispersal ability by smaller organisms, which may result in high distributions globally.

Natural Lake Lifecycles

Lake Creation

Lakes can be formed in a variety of ways, but the most common are discussed briefly below. The oldest and largest systems are the result of tectonic activities. The rift lakes in Africa, for example are the result of seismic activity along the site of separation of two tectonic plates. Ice-formed lakes are created when glaciers recede, leaving behind abnormalities in the landscape shape that are then filled with water. Finally, oxbow lakes are fluvial in origin, resulting when a meandering river bend is pinched off from the main channel.

Natural Extinction

All lakes and ponds receive sediment inputs. Since these systems are not really expanding, it is logical to assume that they will become increasingly shallower in depth, eventually becoming wetlands or terrestrial vegetation. The length of this process should depend upon a combination of depth and sedimentation rate. Moss gives the example of Lake Tanganyika, which reaches a depth of 1500 m and has a sedimentation rate of 0.5 mm/yr. Assuming that sedimentation is not influenced by anthropogenic factors, this system should go extinct in approximately 3 million years. Shallow lentic systems might also fill in as swamps encroach inward from the edges. These processes operate on a much shorter timescale, taking hundreds to thousands of years to complete the extinction process.

Human Impacts

Acidification

Sulfur dioxide and nitrogen oxides are naturally released from volcanoes, organic compounds in the soil, wetlands, and marine systems, but the majority of these compounds come from the combustion of coal, oil, gasoline, and the smelting of ores containing sulfur. These substances dissolve in atmospheric moisture and enter lentic systems as acid rain. Lakes and ponds that contain bedrock that is rich in carbonates have a natural buffer, resulting in no alteration of pH. Systems without this bedrock, however, are very sensitive to acid inputs because they have a low neutralizing capacity, resulting in pH declines even with only small inputs of acid. At a pH of 5–6 algal species diversity and biomass decrease considerably, leading to an increase in water transparency – a characteristic feature of acidified lakes. As the pH continues lower, all fauna becomes less diverse. The most significant feature is the disruption of fish reproduction. Thus, the population is eventually composed of few, old individuals that eventually die and leave the systems without fishes. Acid rain has been especially harmful to lakes in Scandinavia, western Scotland, west Wales and the north eastern United States.

Eutrophication

Eutrophic systems contain a high concentration of phosphorus (~30 μg/L), nitrogen (~1500 μg/L), or both. Phosphorus enters lentic waters from sewage treatment effluents, discharge from raw sewage, or from runoff of farmland. Nitrogen mostly comes from agricultural fertilizers from runoff or leaching and subsequent groundwater flow. This increase in nutrients required for primary producers results in a massive increase of phytoplankton growth, termed a plankton bloom. This bloom decreases water transparency, leading to the loss of submerged plants. The resultant reduction in habitat structure has negative impacts on the species' that utilize it for spawning, maturation and general survival. Additionally, the large number of short-lived phytoplankton results in a massive amount of dead biomass settling into the sediment. Bacteria need large amounts of oxygen to decompose this material, reducing the oxygen concentration of the water. This is especially pronounced in stratified lakes when the thermocline prevents oxygen rich water from the surface to mix with lower levels. Low or anoxic conditions preclude the existence of many taxa that are not physiologically tolerant of these conditions.

Invasive Species

Invasive species have been introduced to lentic systems through both purposeful events (e.g. stocking game and food species) as well as unintentional events (e.g. in ballast water). These organisms can affect natives via competition for prey or habitat, predation, habitat alteration, hybridization, or the introduction of harmful diseases and parasites. With regard to native species, invaders may cause changes in size and age structure, distribution, density, population growth, and may even drive populations to extinction. Examples of prominent invaders of lentic systems include the zebra mussel and sea lamprey in the Great Lakes.

Lotic

Lotic water systems are bodies of water that is flowing continuously, ranging from torrential rapids to slow moving back waters. These systems of water tend to be much shallower than their counterparts, causing temperature to become a major abiotic factor for life in these systems. Water found in these systems will freeze much quicker, and thaw much faster than the deep waters of lentic systems. Lotic ecosystems depend on precipitation, snow melt, and springs to keep the water flowing, in times of drought these shallow systems will dry up and many organisms will die.

Characteristic of Lotic Environment

Moving waters differ in the three major aspects from lakes and ponds : current is a controlling and limiting factor land-water interchange is great because of the small size and depth of moving water systems as compared with lakes; is almost abundant supply except when there is pollution.

Temperature extremes tend to be greater than in standing water. The most distinctive features of moving water ecosystems are related to their motion, i.e., the rate of flow and the stream velocity.

The rate of flow or discharge refers to the volume water observation point during a specific unit of time; in units such as m.3/sec., ft.3/sec., or acre-feet/sec., (one a foot is equivalent to a volume of water 1 acre in area by 1 foot deep). It tends to increase steadily growing downstream, as tributaries join with the main river.

The velocity of flow is the speed at which the water moves, and is measured in m. /sec., ft. / sec., or mi. /hr. Velocity is variable, but it also tends to increase downstream with increasing discharge. A factor that may be even more important to the biotic community is the turbulence, or the irregularity of the motion of the wafer particles.

Perfectly even flow in which water particles move parallel to one another, is -ailed laminar flow it is contrasted with turbulent flow in which the movement of water particles is highly irregular. In highly turbulent water the erosive power is great, the sheer forces at the water-sediment interface at the bottom of the river are powerful, and the amount of oxygen incorporated into the water is very high.

In streams with laminar flow the erosive power is lower, the sheer forces at the bottom of the stream are lower, and relatively less oxygen is incorporated into the water than turbulent waters.

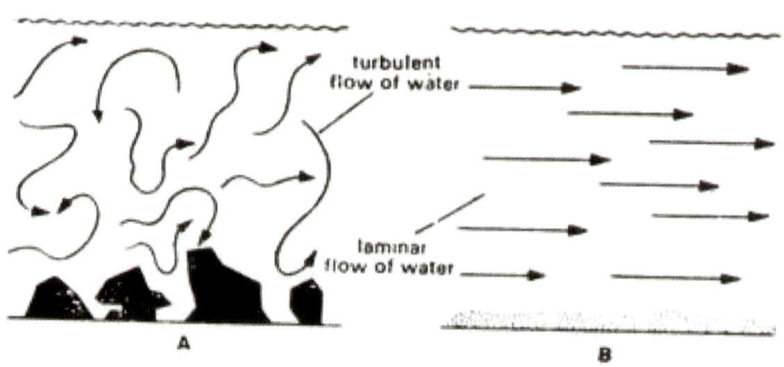

Fig: Diagram showing the differences between turbulent
(A) and laminar (B) flow of water in a stream. The moving-water ecosystems can be divided into following two ecosystems depending up on the velocity and rate of flow of water body.

Rapidly Flowing Water

Rapidly flowing water (of fast moving torrential streams, etc.) can be defined as the portion of the stream in which the flow is both rapid and turbulent. Everything that is not attached or weighty is swept away by the current; this includes organisms and sediment particles alike. The substrate tends to be rock or gravel, and the fragments are rounded and smoothed by the water.

The habitat itself is an extremely diverse one. Physical parameters such as sheer force and rate of water movement tend to be quiet different on top of a rock fragment, between rock fragments, or beneath rock fragments, and different species can exploit these differences in microhabitats.

Aquatic plants, in addition, provide microhabitats for some torrential animals. The rarely occurring phytoplankton of rapid or riffle zone of stream includes diatoms, blue-green and green algae (e.g., Cladophora, Ulothrix) and water moss (e.g., Fontainalis).

Plant Inhabiting Torrential Forms

Some animals of rapidly flowing water streams live among the mosses and flowering plants like Eriocaulon miserum, Hydrous lichen ides, and Duroea walli- chii. Animals inhabiting these plants have a torpedo-like body which enables them to offer minimum resistance to the current.

Some animals such as funnel mouthed tadpoles of Megalophrys and tippled larvae live entangled in the roots of these plants. They possess devices for fixation and have hydrostatic organs.

Some animals such as Cephalopteryx, Helodes, Phalacrocera, Gammarus, etc., live among leaves and stems of aquatic plants. All these animals possess hook-like structures on their body helping them in anchoring on leaves and stems. The larvae of simulids and chironomids, however, live on the exposed surfaces of plants.

Rock Inhabitant Forms

The animals that live on the tops of exposed rocks have an efficient mechanism for staying in one place; otherwise, they will be swept away. The organisms in this microhabitat are universally flattened. Some, like the fresh-water limpet, Ferrissia, are virtually flat, offering little resistance to the current, are holding themselves in place with a very large powerful foot that extends over almost the entire area of the shell.

Others, such as water penny, the larvae of the riffle beetle, and Baetis larvae, are not only almost flat, but each of their legs possesses hooked claws that enable them to hold onto the substrate more firmly. Larvae of Samarium and chironomids cling to rock top by grappling hooks at the posterior end of the body. Caddishly worms contain both of these structures (claws on legs and hooks at posterior end of body).

The mayfly nymphs Iron and Psephenus attach themselves to the rocks by means of functional pads. Some animals, including fresh-water sponges, actually cement themselves to the surface of the rocks. Others, such as caddish flies, build “houses” cut of sediment or wood fragments, which are then cemented firmly to the rocks. Exposed surfaces of rocks also contain few sessile algae and few plants.

A significant amount of organic detritus is washed into rapid-water ecosystems from upstream and adjacent terrestrial areas, however. This is the source of the most of energy in fast water ecosystems, and is much more important as an energy source then primary production within the stream itself. In consequence, the majority of primary consumers of such a microhabitat are detritus-eaters.

Inhabitant Forms of Spaces between Rocks:

Numerous different kinds of animals live in the spaces between the rocks. Many of them, such as mayfly and stonefly naiads, are flattened and have certain behavioral adaptations to hold them in place. These include tbigmotaxis, by which they cling instinctively to any hard surface such as rock or another insect larva. Kheofaxis, by which the animals orient themselves to face into current and move upstream, is another adaptation of this sort.

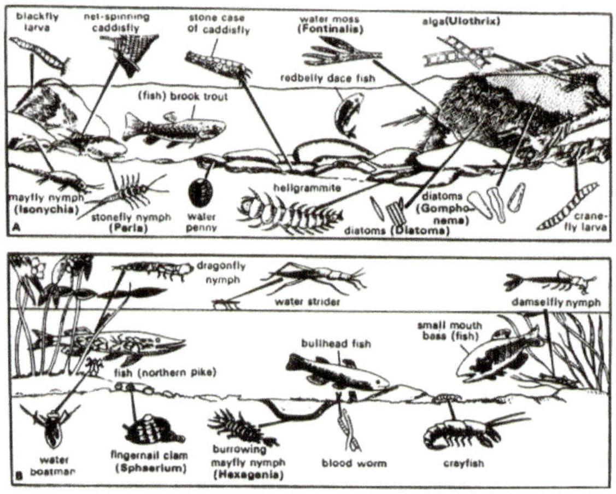

Comparision of biota of rapidly-flowing stream (A) and slowly-flowing stream.

In many of these naiads, the combination of the organisms shape and behaviour is such that the current presses the insect tightly against the rock, increasing the friction between the animal and its substrate. Other insect larvae, such us the hellgrammite, are large and covered with spines their size make it somewhat harder for the current to sweep them away, while the spines help in holding the larvae in places between stones.

Inhabitants Occurring Beneath Rocks

Many species such as flatworms, annelid worms, other insect larvae, clams and some species of snails, live beneath the rocks. The current is weaker here and animals are less likely to be carried away. They lack any special adaptation.

Finally, if the current is sufficiently slow, certain swimming organisms such as fish will be present. The fish of fast-water ecosystems tend to be stenothermal cold-water fish, such as trout. The fishes of fast-water ecosystems have small body size and streamlined body. Trout and masheer are capable of moving against the current or waterfall by muscular effort.

Majority of torrential fishes, however, live on or among rocks and boulders, over which water flows very swiftly. Small loaches (Noemaclicilus) or loach-like fishes (Ambyceps olyra) are met with at the bottom and limpet-like fishes (e.g., Glyptosternum, Balilora) or fishes with special modifications, Garra, Glyptothorax, Pseudecheneis) occur on the rocks.

The loaches and carps adhere to rocks like the limpets, with the ventral surface, inclusive of paired fins forming one large suckcr, e.g., Gastromyzvn of Borneo, the Indian loach Baletora of the eastern Himalayas and Assam hills.

Physico-chemically, the fast-water ecosystems resemble with cold, deep lakes. Water temperatures tend to be quite low, productivity is also quite low, and diversity is high. In the fast-water ecosystem, the main control of productivity is the current, which seriously limits the amount and type of autotrophy production that can take place.

Slowly Flowing Water

A slowly flowing water ecosystem is a very different type of system from the fast streams. Because the flow is both slower and more likely to be laminar, the erosive power of the stream is greatly reduced, and smaller sediment particles (silt) and decaying organic debris, instead of being carried away by the stream, are deposited on the bottom.

In addition to these the slow streams have a higher temperature, consequently, plank tonic organisms, especially protozoans, occur in large numbers in this ecosystem the detritus-feeding benthos of stow water ecosystems include those which either live on the bottom, such asisopods (sow bugs), mollusks (Sphaenius, Pisidiunu and Anodonta dominata), and mayfly and damselfly naiads, or which burrow into the sediment, such as tubeworms, naiads of the burrowing mayflies, Sialis (alderfly), midge Chironomus, and several other insect larvae, as well as, clams, nematodes, snails and rotifers.

Swimming organisms are also abundant, including not only fishes, such as carps, cat fishes, suckers, stingers, spoonbills, etc., which tend to be different species from those of fast-water areas, but also larger Crustacea such as amphipods (fresh-water shrimps).

Finally, several insects spend most of their time at the surface of the stream. These include forms such as water striders, water boatmen, back- swimmers, and predaceous diving beetles.

Zooplankton is abundant, including a rich assemblage of Protozoa and smaller Crustacea such as Cladocera (water-fleas) and copepods. The richness of wood afforded by slow streams attracts a large number of reptiles (water- snakes, crocodiles, turtles) and amphibians.

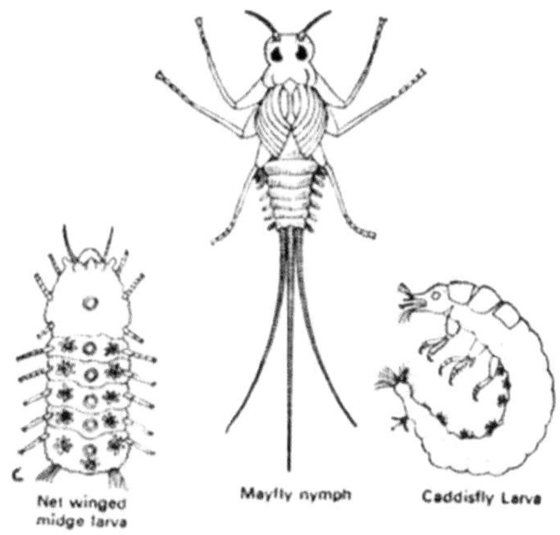

Some inhabitants of lotic environment.

Plant life is also abundant in a slow-water ecosystem. It includes rooted vascular plants such as pond weeds and grasses firmly attached aquatic mosses and large multicellular filamentous algae. Motile algae such as diatoms and flagellates may abound in the open in all, the primary productivity of slow-water ecosystem is than that of the rapid-water ecosystem, and the community is relatively less dependent on food materials from outside.

Further, detritus food chain consisting of organisms such as bacteria t: fungi is much better developed in this community as decomposed organic debris that comprises the main food for these organisms accumulates in the mud bottom some slow-moving streams, in fact, the bottom muds contain more organic material than mineral fragments.

In slow-water streams oxygen concentration is main limiting factor. The high level of animal activity, coupled with an active detritus food chain, can withdraw a large amount of oxygen from water of a slowly moving stream.

In addition, the low level of turbulence means that less oxygen is incorporated into the water at the surface. Thus the dissolved oxygen content of a slowly moving stream is likely to be much lower than that of a fast-moving stream. This is commonly reflected in the fauna of the stream.

Fishes tolerant of low oxygen levels such as carps and catfishes are the most common fishes in slow water, while species with high oxygen demands, such as trout are found in fast water.

Trophic Relationships

Pondweed is an autochthonous energy source.

Leaf litter is an allochthonous energy source.

Energy Inputs

Energy sources can be autochthonous or allochthonous.

- Autochthonous: energy sources are those derived from within the lotic system. During photosynthesis, for example, primary producers form organic carbon compounds out of carbon dioxide and inorganic matter. The energy they produce is important for the community because it may be transferred to higher trophic levels via consumption. Additionally, high rates of primary production can introduce dissolved organic matter (DOM) to the waters. Another form of autochthonous energy comes from the decomposition of dead organisms and feces that originate within the lotic system. In this case, bacteria decompose the detritus or coarse particulate organic material (CPOM; >1 mm pieces) into fine particulate organic matter (FPOM; <1 mm pieces) and then further into inorganic compounds that are required for photosynthesis. This process is discussed in more detail below.
- Allochthonous: energy sources are those derived from outside the lotic system, that is, from the terrestrial environment. Leaves, twigs, fruits, etc. are typical forms of terrestrial CPOM that have entered the water by direct litter fall or lateral leaf blow. In addition, terrestrial animal-derived materials, such as feces or carcasses that have been added to the system are examples of allochthonous CPOM. The CPOM undergoes a specific process of degradation. Allan gives the example of a leaf fallen into a stream. First, the soluble chemicals are dissolved and leached from the leaf upon its saturation with water. This adds to the DOM load in the system. Next, microbes such as bacteria and fungi colonize the leaf, softening it as the mycelium of the fungus grows into it. The composition of the microbial community is influenced by the species of tree from which the leaves are shed (Rubbo and Kiesecker 2004). This combination of bacteria, fungi, and leaf are a food source for shredding invertebrates, which leave only FPOM after consumption. These fine particles may be colonized by microbes again or serve as a food source for animals that consume FPOM. Organic matter can also enter the lotic system already in the FPOM stage by wind, surface runoff, bank erosion, or groundwater. Similarly, DOM can be introduced through canopy drip from rain or from surface flows.

Invertebrates

Invertebrates can be organized into many feeding guilds in lotic systems. Some species are shredders, which use large and powerful mouth parts to feed on non-woody CPOM and their associated microorganisms. Others are suspension feeders, which use their setae, filtering aparati, nets, or even secretions to collect FPOM and microbes from the water. These species may be passive collectors, utilizing the natural flow of the system, or they may generate their own current to draw water, and also, FPOM in Allan. Members of the gatherer-collector guild actively search for FPOM under rocks and in other places where the stream flow has slackened enough to allow deposition. Grazing invertebrates utilize scraping, rasping, and browsing adaptations to feed on periphyton and detritus. Finally, several families are predatory, capturing and consuming animal prey. Both the number of species and the abundance of individuals within each guild is largely dependent upon food availability. Thus, these values may vary across both seasons and systems.

Fish

Fish can also be placed into feeding guilds. Planktivores pick plankton out of the water column. Herbivore-detritivores are bottom-feeding species that ingest both periphyton and detritus indiscriminately. Surface and water column feeders capture surface prey (mainly terrestrial and emerging insects) and drift (benthic invertebrates floating downstream). Benthic invertebrate feeders prey primarily on immature insects, but will also consume other benthic invertebrates. Top predators consume fishes and/or large invertebrates. Omnivores ingest a wide range of prey. These can be floral, faunal, and/or detrital in nature. Finally, parasites live off of host species, typically other fishes. Fish are flexible in their feeding roles, capturing different prey with regard to seasonal availability and their own developmental stage. Thus, they may occupy multiple feeding guilds in their lifetime. The number of species in each guild can vary greatly between systems, with temperate warm water streams having the most benthic invertebrate feeders, and tropical systems having large numbers of detritus feeders due to high rates of allochthonous input.

Iguazu Falls: an extreme lotic environment

Beaver Run: a placid lotic environment

Community Patterns and Diversity

Local Species Richness

Large rivers have comparatively more species than small streams. Many relate this pattern to the greater area and volume of larger systems, as well as an increase in habitat diversity. Some systems, however, show a poor fit between system size and species richness. In these cases, a combination of factors such as historical rates of speciation and extinction, type of substrate, microhabitat availability, water chemistry, temperature, and disturbance such as flooding seem to be important.

Resource Partitioning

Although many alternate theories have been postulated for the ability of guild-mates to coexist, resource partitioning has been well documented in lotic systems as a means of reducing competition. The three main types of resource partitioning include habitat, dietary, and temporal segregation.

Habitat segregation was found to be the most common type of resource partitioning in natural systems. In lotic systems, microhabitats provide a level of physical complexity that can support a diverse array of organisms. The separation of species by substrate preferences has been well documented for invertebrates. Ward was able to divide substrate dwellers into six broad assemblages, including those that live in: coarse substrate, gravel, sand, mud, woody debris, and those associated with plants, showing one layer of segregation. On a smaller scale, further habitat partitioning can occur on or around a single substrate, such as a piece of gravel. Some invertebrates prefer the high flow areas on the exposed top of the gravel, while others reside in the crevices between one piece of gravel and the next, while still others live on the bottom of this gravel piece.

Dietary segregation is the second-most common type of resource partitioning. High degrees of morphological specializations or behavioral differences allow organisms to use specific resources. The size of nets built by some species of invertebrate suspension feeders, for example, can filter varying particle size of FPOM from the water. Similarly, members in the grazing guild can specialize in the harvesting of algae or detritus depending upon the morphology of their scraping apparatus. In addition, certain species seem to show a preference for specific algal species.

Temporal segregation is a less common form of resource partitioning, but it is nonetheless an observed phenomenon. Typically, it accounts for coexistence by relating it to differences in life history patterns and the timing of maximum growth among guild mates. Tropical fishes in Borneo, for example, have shifted to shorter life spans in response to the ecological niche reduction felt with increasing levels of species richness in their ecosystem.

Persistence and Succession

Over long time scales, there is a tendency for species composition in pristine systems to remain in a stable state. This has been found for both invertebrate and fish species. On shorter time scales, however, flow variability and unusual precipitation patterns decrease habitat stability and can all lead to declines in persistence levels. The ability to maintain this persistence over long time scales is related to the ability of lotic systems to return to the original community configuration relatively quickly after a disturbance. This is one example of temporal succession, a site-specific change in a community involving changes in species composition over time. Another form of temporal succession might occur when a new habitat is opened up for colonization. In these cases, an entirely new community that is well adapted to the conditions found in this new area can establish itself.

Meandering stream in Waitomo, New Zealand

River Gryffe in Scotland

Rocky stream in Hawaii

River Continuum Concept

The River continuum concept (RCC) was an attempt to construct a single framework to describe the function of temperate lotic ecosystems from the source to the end and relate it to changes in the biotic community. The physical basis for RCC is size and location along the gradient from a small stream eventually linked to a large river. Stream order is used as the physical measure of the position along the RCC.

According to the RCC, low ordered sites are small shaded streams where allochthonous inputs of CPOM are a necessary resource for consumers. As the river widens at mid-ordered sites, energy inputs should change. Ample sunlight should reach the bottom in these systems to support significant periphyton production. Additionally, the biological processing of CPOM (Coarse Particulate Organic Matter - larger than 1 mm) inputs at upstream sites is expected to result in the transport of large amounts of FPOM (Fine Particulate Organic Matter - smaller than 1 mm) to these downstream ecosystems. Plants should become more abundant at edges of the river with increasing river size, especially in lowland rivers where finer sediments have been deposited and facilitate rooting. The main channels likely have too much current and turbidity and a lack of substrate to support plants or periphyton. Phytoplankton should produce the only autochthonous inputs here, but photosynthetic rates will be limited due to turbidity and mixing. Thus, allochthonous inputs are expected to be the primary energy source for large rivers. This FPOM will come from both upstream sites via the decomposition process and through lateral inputs from floodplains.

Biota should change with this change in energy from the headwaters to the mouth of these systems. Namely, shredders should prosper in low-ordered systems and grazers in mid-ordered sites. Microbial decomposition should play the largest role in energy production for low-ordered sites and large rivers, while photosynthesis, in addition to degraded allochthonous inputs from upstream will be essential in mid-ordered systems. As mid-ordered sites will theoretically receive the largest variety of energy inputs, they might be expected to host the most biological diversity.

Just how well the RCC actually reflects patterns in natural systems is uncertain and its generality can be a handicap when applied to diverse and specific situations. The most noted criticisms of the RCC are:

1. It focuses mostly on macro invertebrates, disregarding that plankton and fish diversity is highest in high orders;
2. It relies heavily on the fact that low ordered sites have high CPOM inputs, even though many streams lack riparian habitats;
3. It is based on pristine systems, which rarely exist today; and,
4. It is centered around the functioning of temperate streams. Despite its shortcomings, the RCC remains a useful idea for describing how the patterns of ecological functions in a lotic system can vary from the source to the mouth.

Disturbances such as congestion by dams or natural events such as shore flooding are not included in the RCC model. Various researchers have since expanded the model to account for such irregularities. For example, J.V. Ward and J.A. Stanford came up with the Serial

Discontinuity Concept in 1983, which addresses the impact of geomorphologic disorders such as congestion and integrated inflows. The same authors presented the Hyporheic Corridor concept in 1993, in which the vertical (in depth) and lateral (from shore to shore) structural complexity of the river were connected. The flood pulse concept, developed by W.J. Junk in 1989, further modified by P.B. Bayley in 1990 and K. Tockner in 2000, takes into account the large amount of nutrients and organic material that makes its way into a river from the sediment of surrounding flooded land.

Human Impacts

Pollution

Pollutant sources of lotic systems are hard to control because they derive, often in small amounts, over a very wide area and enter the system at many locations along its length. Agricultural fields often deliver large quantities of sediments, nutrients, and chemicals to nearby streams and rivers. Urban and residential areas can also add to this pollution when contaminants are accumulated on impervious surfaces such as roads and parking lots that then drain into the system. Elevated nutrient concentrations, especially nitrogen and phosphorus which are key components of fertilizers, can increase periphyton growth, which can be particularly dangerous in slow-moving streams. Another pollutant, acid rain, forms from sulfur dioxide and nitrous oxide emitted from factories and power stations. These substances readily dissolve in atmospheric moisture and enter lotic systems through precipitation. This can lower the pH of these sites, affecting all trophic levels from algae to vertebrates. Mean species richness and total species numbers within a system decrease with decreasing pH.

While direct pollution of lotic systems has been greatly reduced in the United States under the government's Clean Water Act, contaminants from diffuse non-point sources remain a large problem.

Flow Modification

A weir on the River Calder, West Yorkshire.

Dams alter the flow, temperature, and sediment regime of lotic systems. Additionally, many rivers are dammed at multiple locations, amplifying the impact. Dams can cause enhanced clarity and reduced variability in stream flow, which in turn cause an increase in periphyton abundance.

Invertebrates immediately below a dam can show reductions in species richness due to an overall reduction in habitat heterogeneity. Also, thermal changes can affect insect development, with abnormally warm winter temperatures obscuring cues to break egg diapause and overly cool summer temperatures leaving too few acceptable days to complete growth. Finally, dams fragment river systems, isolating previously continuous populations, and preventing the migrations of anadromous and catadromous species.

Invasive Species

Invasive species have been introduced to lotic systems through both purposeful events (e.g. stocking game and food species) as well as unintentional events (e.g. hitchhikers on boats or fishing waders). These organisms can affect natives via competition for prey or habitat, predation, habitat alteration, hybridization, or the introduction of harmful diseases and parasites. Once established, these species can be difficult to control or eradicate, particularly because of the connectivity of lotic systems. Invasive species can be especially harmful in areas that have endangered biota, such as mussels in the Southeast United States, or those that have localized endemic species, like lotic systems west of the Rocky Mountains, where many species evolved in isolation.

Wetlands

Wetland is a complex ecosystem characterized by flooding or saturation of the soil, which creates low-oxygen environments that favour a specialized assemblage of plants, animals, and microbes, which exhibit adaptations designed to tolerate periods of sluggishly moving or standing water. Wetlands are usually classified according to soil and plant life as bogs, marshes, swamps, and other similar environments.

Wetlands and the sub discipline of wetland ecology are a relatively new area of study in the field of ecology, primarily arising out of the laws and other regulations enacted during the 1970s. The term wetland, however, was first used formally in 1953, in a report by the U.S. Fish and Wildlife Service (USFWS) that provided a framework for a later publication concerning waterfowl habitat in the United States. Since then, wetlands have been variously defined by ecologists and government officials. No single, formal definition exists; however, the definition provided by the Ramsar Convention, an intergovernmental treaty signed in Ramsar, Iran, in 1971 to guide national and international wetland-conservation measures, is among the most widely referenced:

Wetlands are areas of marsh, fen, peatland or water, whether natural or artificial, permanent or temporary, with water that is static or flowing, fresh, brackish or salt, including areas of marine water the depth of which at low tide does not exceed six metres.

This definition is also broad enough to encompass open water used by birds—the concept that originally inspired the protection of wetlands and associated aquatic sites.

Evidence of the first wetland plants extends back to the Ordovician Period (485.4 million to 443.8 million years ago), when the first terrestrial plants, which were dependent on wet substrates, began to colonize the land. Wetland plants and the animals that depended on them continued to evolve, and the first marshes and swamps appeared during the Devonian Period (419.2 million to 358.9 million years ago). Swamps later dominated vast regions, such as the land that would become southern North America, during the Carboniferous Period (358.9 million to 298.9 million years ago), and parts of the Mesozoic and Cenozoic eras (252.2 million years ago to the present) were also characterized by the presence of extensive wetland habitats.

prairie pothole: Prairie potholes are unusual in that they are fed mostly by precipitation or groundwater and are not associated with a distinct body of water, such as a river or a larger lake.

Wetland communities depend on access to liquid water. Throughout geologic history, water availability has varied according to prevailing local and global climate patterns, latitude, elevation, season, and distance from both water bodies and groundwater. As a result of this variability, wetland communities in different parts of the world are the product of different conditions.

Glaciation during the Pleistocene Epoch (2.6 million to 11,700 years ago) prepared several types of landscapes for the development of present-day wetlands. In glaciated regions, the movement of ice sheets scoured the landscape, and the weight of the ice depressed Earth's crust below. Both processes created low-relief areas, such as the flat, scoured landscape of Canada's Hudson Bay lowlands. This region, which hosts extensive wetlands that are fed by groundwater and precipitation, continues to experience isostatic uplift (a rebound in the land that follows a glacier's retreat) that brings more of Hudson Bay's bottom to the surface. Some of this new land has become vegetated, and the wetlands have expanded. As the Pleistocene glaciers retreated across the Northern Hemisphere, melt carved wide, flat valleys that are occupied today by major rivers and their associated wetlands and floodplains (flat land area adjacent to a stream). Uneven scouring of the landscape in some regions resulted in low spots that filled with melted snow and rainwater during particularly wet years. This process created the prairie pothole region of the Midwest and south-central Canada.

In some of the coldest parts of the world, wetlands are sustained by an impermeable layer of ice that remains in the soil throughout the year. This perennially frozen ground, or permafrost, prevents both the percolation of surface water into the ground and plant contact with mineral groundwater. About 20–22 percent of Earth's land surface is close enough to a polar region or high enough in altitude to experience permafrost. Much of northern North America and Eurasia, as well as the Mongolian and Tibetan Plateaus, are affected by permafrost, and these regions host vast expanses of bogs, fens, and peat lands. North America possesses some of the most extensive bog and fen regions on Earth. In western Siberia, larch-spruce-birch forests form part of an enormous inland delta, which is the largest contiguous area of peat lands in the world. Asian plateaus in general host some of the most unusual high-altitude wetland ecosystems.

Some wetland areas were created in other ways during periods of low sea level, when water was locked in glacial ice. Following a drop in sea level, the coastal plain of the southeastern United States was formed by the deposition of sediment that eroded from landscapes upstream. Rising sea levels that followed the retreat of the glaciers reduced stream flow velocity, and many streams backed up. These changes resulted in the formation of a variety of depressional, flat, and riverine wetlands. In addition, the warming period that directly followed the most recent glacial episode (which ended approximately 11,700 years ago) was marked by rivers flowing with melted ice water, buried chunks of ice that melted and formed kettle lakes, large lakes that formed in low areas inland, sea margins that moved inland, and coastal water tables that generally rose with the sea. Wetlands subsequently developed along lake and coastal margins, in delta areas, and across floodplains.

Wetlands in non-glaciated regions, such as the tropics, were developed during periods characterized by slightly different climates and thus may be changing under present-day conditions. The peat swamp forests of Indonesia are built on peat up to 15 metres (about 50 feet) thick. The accumulation of this material occurred during a wetter period several thousand years ago. Although new peat is still forming in places and the region remains humid, the tropical climate of the present day is dry enough to allow the degradation of peat in some areas.

Okavango delta, northern Botswana.

Geographic Distribution of Wetlands

Wetlands are found all over the world in every biome, or major life zone. Some wetlands, such as tidal marshes, fit the definition of a transitional zone because they occur where open water and land meet. Others, such as prairie potholes of central North America and Carolina bays (elliptical depressions) of the Atlantic Coastal Plain, are fed mostly by precipitation or groundwater and are not associated with a distinct body of water. Still others, such as the bog-and-fen mosaic of the taiga (boreal forest), are dominant features of the regional landscape.

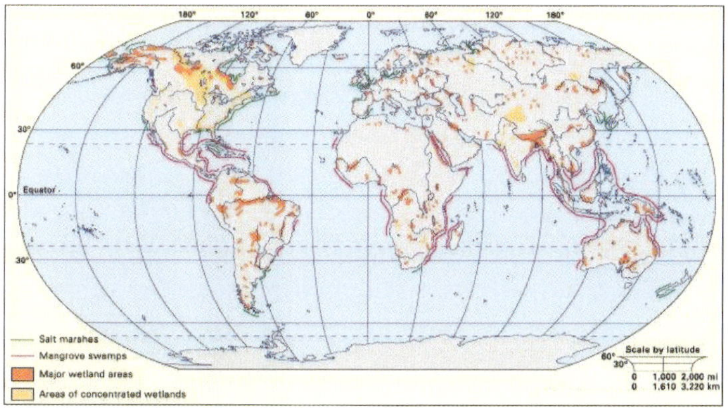

Wetland: Major wetland areas and worldwide distribution of salt marshes and mangrove swamps.

Wetlands are most abundant in boreal and tropical regions, though a wide variety of inland and coastal wetlands are also found in temperate regions. This distribution is generally due to conditions that promote an abundance of water. For example, the peat lands of Borneo and the Peruvian Amazon occur within the tropical rainforest biome. In the treeless tundra of Alaska and Canada, saturated and flooded wetlands are underlain by permafrost. There, many parts of the landscape are made up of peat lands composed of black spruce (Picea Mariana) and white spruce (P. glauca), which are sustained by rainfall or melting snow. Wetlands are also found in the hot desert biome—for example, the Mesopotamian marshlands found at the confluence of the Tigris and Euphrates rivers. In temperate regions, wetlands typically are found near coastlines, rivers, lakes, or other locations where local water input exceeds output.

Environmental Conditions

Climate

Wetland formation is influenced by climate patterns and the limitations posed by landforms. The net balance of precipitation and evaporation determines the quantity and timing of water available for the formation or maintenance of wetland conditions. Water flows downhill, and the geomorphology of the landscape determines where it gathers as well as what topography or subsurface layers prevent it from draining away. Each wetland has a water signature, or hydro period, which is characterized by the timing, duration, and quantity of water in the system. Furthermore, inundation has biological consequences, because it prevents atmospheric oxygen from being replenished in the soil. As a result, only organisms that can tolerate or are adapted to low-oxygen or anoxic (negligible oxygen) conditions have an advantage in wetland environments.

Soils

Wetland, or hydric, soils form when saturated or flooded conditions last long enough during the growing season to cause anaerobic (oxygen-depleted) regions to occur in the upper part of the soil, which includes the root zone. Such soils can be organic (containing organic compounds) or derived from minerals. Organic wetland soils, such as peatland soils, contain at least 12 percent organic matter and are typically acidic; they also possess a high water-holding capacity and low nutrient availability. Organic matter builds up in the soil when low oxygen conditions halt or slow decomposition. Mineral wetland soils, on the other hand, have less than 12 percent organic matter, and they often exhibit gleying, where ferric iron (Fe^{3+}) and manganese are reduced (that is, they gain electrons) in the soil by anaerobic bacteria thriving in the depleted oxygen conditions. The resulting ferrous iron (Fe^{2+}) becomes concentrated in a deep soil layer (soil horizon). In waterlogged soils, the topsoil and upper soil layers take on a black, gray, or blue-green colour. Pore linings (the coatings on the surface of open spaces in the soil) in wetland mineral soils are often red, because plant roots, which make many of the pores, release oxygen into the oxygen-depleted soil. The presence of this oxygen in an anaerobic environment oxidizes some of the ferrous iron remaining in the water and concentrates it along the pore linings. Plant litter and animal waste may occur at the surface of flooded mineral soils.

peat bed: Partially excavated peat bed near Avon Park, Florida.

Wetland Types

Various classification systems of wetlands have been developed to serve different purposes. In *Classification of Wetlands and Deep water Habitats of the United States* (1979), the USFWS presented a hierarchical system based on five ecosystem types: marine, estuarine, riverine, lacustrine, and palustrine. Similarly, the Ramsar Convention based its classification system on the USFWS model, but it added a human-created, or cultural, wetland type. Other classification systems are more consistently based on structure (i.e., the physical appearance of the wetlands), function (i.e., the hydrologic regimes and the role of the wetlands within them), or management goals (i.e., how the wetlands are used by humans).

People worldwide have long applied unique terminology for the wet places in their landscapes; however, this terminology has not been standardized across all classifications. The categories presented in the following sections are based not on a formal classification system but rather on general terms in common use that cover most wetlands.

Coastal Systems

Mangroves

Mangroves are found in tropical and subtropical coastal areas between 32 °N and 38 °S. They are sensitive to cold temperatures and are generally found in regions that do not experience hard frosts (the average air temperature of the coldest months does not fall below 20 °C [68 °F]). Some mangroves, however, may survive drops in temperature to 7 °C (about 45 °F) or lower. Mangrove populations located farthest from the Equator are found in regions warmed by equatorial currents or in somewhat fresher estuarine waters, where reduced salt stress may enable the trees to handle colder temperatures.

Silver baitfish in a mangrove channel on South island, Bahamas.

Mangrove seeds are dispersed by water currents, and habitats that are favourable for mangrove growth are somewhat protected from wave action, where lower-energy conditions allow mangrove seedlings to establish. The largest expanses of mangroves occur in wet deltaic regions, such as the Sundarbans on the Ganges delta in India and Bangladesh, the Niger Delta complex in Nigeria and Cameroon, and the Orinoco and Gulf of Paria deltas on Venezuela's east coast. Other coastlines hosting nearly continuous stands of mangroves include northern Brazil, the deltaic coast of southern Papua New Guinea, and the West African coast from southern Senegal to central Sierra Leone.

Salt Marshes

Salt marshes are found primarily along temperate and some boreal coastlines where sediment accumulates and mangroves do not dominate; however, some also occur in Arctic and tropical regions. Salt marshes flourish wherever the accumulation of sediments is equal to or greater than the rate of land subsidence and where there is adequate protection from disruptive waves and storms. Some inland wetland systems develop saline conditions when the rate of evapotranspiration (combined water loss from evaporation from the soil and transpiration from plants) exceeds the rate of precipitation or as a result of contact with saline groundwater. In areas characterized by the presence of saline groundwater, such as eastern Nebraska, vegetation similar to that of coastal salt marshes is found.

Salt marsh in East Lyme, Connecticut.

Tidal Freshwater Marshes

Tidal freshwater marshes are found in large river systems throughout the world from subarctic regions to the Equator. The largest expanses of these wetlands occur at temperate latitudes. Tidal freshwater marshes are found in segments of river systems that are close enough to the coast to experience significant tidal action but not salinity. Downstream, the mouths of these rivers experience some tidal flux, and narrowing river channels upstream constrict the volume of the incoming tide so that the overall change in water depth is greater upriver than at the coast.

This marsh type is found extensively along the eastern and Gulf coasts of North America, as well as in the large river basins on the west coast. Other large river systems with extensive tidal freshwater marshes include the St. Lawrence River between the United States and Canada, the Rhine and Thames rivers in Europe, and the Yellow River (Huang He) in Asia.

Intertidal Flats

Intertidal flats are relatively flat unvegetated areas between tidal marshes and deeper water. They are found in somewhat protected areas with limited wave action that are nevertheless disturbed by wind, waves, and currents. Their sediments are too unstable and the physical energy and the duration of flooding are too great for communities of large plants to become established. Intertidal flats are found throughout the world, such as in the St. Lawrence River lowlands, in areas adjacent to the sea grass beds of coastal Mexico, and in Hudson Bay, Canada.

Inland Systems

Freshwater Marshes

The wetlands in this diverse group are unified primarily by the fact that they are all nontidal, non-forested freshwater systems dominated by grasses, sedges, and other freshwater hydrophytes (aquatic plants). Freshwater marshes do not build peat. They differ in their geologic origins and their driving hydrologic forces, and they vary in size from small pothole marshes less than a hectare in size to the immense expanses of saw grass, such as those that are found in the Florida Everglades.

A freshwater marsh, dominated by saw grass and dotted by palms and cypresses, in the Everglades, southern Florida.

Freshwater marshes occur worldwide in low areas of the landscape where water collects or where a relatively impermeable soil or geologic layer causes water to pond. This group of wetlands includes the prairie potholes, vernal pools, and playas of North America, as well as the vegetated fringes of small lakes, the coastal lagoons behind the beaches of barrier islands, and the delta marshes of tributary rivers of enormous lakes (e.g., the Great Lakes of North America). The group also includes marsh systems that deliver strong seasonal pulses of fresh water to more saline areas, a dynamic illustrated by the Everglades of south Florida, where during the wet season a sheet of fresh water flows from Lake Okeechobee to the ocean.

Peatlands

Peatlands (which are sometimes called moors in Europe; they may also be referred to as mires when they are actively forming peat) develop in areas where conditions cause plant material to decompose so slowly that there is a net accumulation of organic matter (peat) each growing season. Two types of peatlands, bogs and fens, have been studied extensively in high latitudes. Bogs develop in depressions that are low in nutrients and fed primarily by rainfall, whereas fens develop on slopes, in depressions, or on flats as a result of sustained flows of mineral-rich groundwater in the root zone. Over time, bog or fen patches may merge to form a blanket over a broad area.

Bogs and fens are found extensively in the cool and moist boreal regions of the Northern Hemisphere, where evaporation is low and moisture accumulates from ample precipitation and high humidity from maritime influences. The landscapes of Canada that were once overlain by glaciers host the largest peatlands (about 1.1 million square km [425,000 square miles]) in the world. Scandinavia, Eastern Europe, western Siberia, and Alaska harbour the balance of peatlands in the cold temperate region. In the United States, bogs and fens are found primarily in clusters around the Great Lakes and in Maine; these peatlands usually develop in basins that were scoured out by glaciers during the Pleistocene Epoch.

Overijssel: peat bog near Enschede, Overijssel province, Netherlands.

Lower-latitude peatlands also exist. Pocosins (or evergreen shrub bogs) are well-documented peatlands of the south-eastern United States, which has a humid subtropical climate. Within that region, pocosins occur mainly on the flat plateaus of the Middle Atlantic Coastal Plain, especially in North Carolina, where waterlogged, acidic, and nutrient-poor soils are made up of a mixture of sand and peat. Pocosins may be dominated by shrubs of the heath family (Ericaceae) and pines or conifers and hardwoods.

Tropical peatlands are found in the lowlands of East Asia and Southeast Asia, the Caribbean, Central and South America, and Africa. The largest known expanses of these peatlands occur in Indonesia, where they cover 10–12 percent of the country. Some of these peatlands are tropical bogs, which occur in Southeast Asia and tropical parts of Africa and South America, but they are not as well studied as those that occur in higher latitudes.

Less well known are tropical peat swamps, which form in floodplains and tropical domed bogs (vegetation-covered mounds of peat). Tropical bog trees, such as Calophyllum and Shorea, often grow on hummocks situated above the water table, because their seeds often require drier, more-aerated conditions in order to germinate and become established. Hummocks are sometimes made up of trunks of old dead trees. Many domed bog trees are equipped with stilt roots, which grow above the water surface and act as flying buttresses to support the tree, or with pneumatophores (upward-growing structures connected to the plant's roots), which project into the air and transport atmospheric oxygen to roots, some of which extend up to 2 metres (6.6 feet) deep in the anoxic zone.

The tropical domed peatlands of Borneo host an astonishing diversity of woody plants. Calophyllum, Combretocarpus, and Cratoxylum often appear together in some wetter evergreen forests. In contrast, Dactylocladus, Gonystylus, and Shorea are found together in peat swamps. Narrow-leafed palm like pandans (Pandanus and Freycinetia) cover the ground in "low-pole" (or Padang) forests where light penetrates canopy gaps. These lowland forests of Borneo provide habitat for the endangered orangutans of genus Pongo.

Tropical peat swamp forests develop where conditions create permanently saturated acidic substrates. Their formation is a complex process, and peat accumulation in a given location can vary with changes in climate, along with river configuration and flow. Some tropical peatlands formed during wetter periods and may now be experiencing peat degradation due to the relatively drier climate of the present day.

Although less widespread in the Southern Hemisphere, peatlands are also found in the subalpine zones of the Patagonian Andes in South America, as well as in the lowlands of New Zealand. Patagonia's peatlands are characterized by species of sphagnum moss (Sphagnum), sedge (Carex), rush (Juncus), and grasses (Agrostis). New Zealand's peatlands occur in raised blanket bogs (or flat elevated bogs) dominated by restiads (wire rushes), such as Sporadanthus and Empodisma.

Freshwater Forested Swamps

A man canoes among cypress trees in the Okefenokee Swamp, in south-eastern Georgia.

Freshwater forested swamps are dominated by trees or other woody vegetation. These wetland systems occur throughout the world. In the tropics, vast swamps are found along the great rivers, by which they are often inundated for many months. The entire Atlantic coast of the temperate United States is particularly rich with swamps.

Freshwater forested swamps characterized by red maple (*Acer rubrum*) are found in the formerly glaciated north-eastern United States. Coastal swamps, which are found from the coast of Maine to the Gulf Coast, are dominated by Atlantic white cedar (*Chamaecyparis thyoides*), however. Atlantic white cedar swamps, which occupy sites that are drier than deep water swamps, are flooded in winter and for an extended period during spring. Red maple swamps, in contrast, experience less flooding.

Deep water swamps characterized by cypress (*Cupressus*) and tupelo (*Nyassa*) trees are found from Delaware to Texas and along the Mississippi River, extending north to Illinois. However, they occur primarily along the wide meandering rivers of the Atlantic Coast Plain. On this plain, the flat topography, which slows the emptying of rivers into the ocean, combined with rising ocean levels that followed the retreat of the glaciers to create favourably wet conditions. In general, deep water cypress-tupelo swamps are inundated 90–100 percent of the year, with low- to moderate-strength currents.

Back swamps—basins occurring behind the natural levies of a floodplain that are composed of fine flood-deposited sediments—form in abandoned channels (oxbows) or elongated sloughs. Both of these landforms are permanently inundated with water but receive a pulse of nutrient-rich river water and sediments only during the flood season.

Riparian Wetlands

Riparian wetlands are also called riverine wetlands or floodplain wetlands, and they constitute a subset of the riparian system. Riparian systems are linear and open. Not only do they interact with upstream and downstream channels; they also interact laterally with aquatic and terrestrial systems on either side. Riparian wetlands occur along rivers and streams that periodically overflow their channel confines, causing flooding to which the wetland vegetation is adapted. They are also found where a meandering stream channel creates new sites for plant life to take root and grow.

Riparian wetlands are found the world over and take different forms in different regions in response to climatic and topographic factors. Riparian ecosystems can exist as broad and nearly flat alluvial valleys, such as those that occur in the Amazon Basin of South America, in Bangladesh, and in the floodplains of large rivers such as the Mississippi in the United States. In arid regions, however, riparian ecosystems can be narrow strips of vegetation along the bank of a stream that is prone to flash flooding and unstable. In mountainous regions, such as the Pacific Northwest, riparian systems may be narrow along steep headwater streams, but they may widen to expansive floodplains in the lowland reaches.

In arid regions, riparian systems can exist within the valley floors and floodplains of perennial streams and alongside or within ephemeral streams. In most nonarid regions, riparian zones usually develop first along the region of the stream where water flow is constant (where sufficient groundwater enters the channel to sustain flow through dry periods). Many alluvial forests, which form within alluvial fans (fan-shaped deposits of unconsolidated sedimentary material at the mouth of a mountain canyon), may be inundated for part of the year.

Wetland Functions and Ecosystem Benefits

Wetland functions are defined as the physical, chemical, and biological processes or attributes that are vital to the integrity of the wetland system. Because wetlands are often transition zones (ecotones) between terrestrial and deep water aquatic systems, many processes have major implications for species. Since wetlands may provide food and habitat for many terrestrial and many aquatic species, wetland biodiversity is often higher than that of adjacent ecosystems. In addition, wetlands can affect the export of organic materials and serve as a sink for inorganic nutrients and atmospheric carbon. They play a major role in the biosphere by providing habitats for several plants, animals, and other forms of life; they may also serve as the last refuges for many rare and endangered species.

Some wetlands, such as swamps and marshes, are considered to be some of Earth's most productive ecosystems. To humans, wetlands are valuable for their sport fishing, hunting, and recreational uses. In addition, the capacity of wetlands to absorb a great amount of water also benefits developed areas, especially during periods of flooding. Wetland systems can also protect shorelines, recharge groundwater aquifers, and cleanse polluted waters. They have been described as "the kidneys of the landscape".

Community Structure and Ecosystem Development

Since plants remain rooted in one place, plant composition and community structure change over time. Based on early bog studies, the classic view of lake succession was that a shallow lake would fill in overtime and become a wetland. As the area continued to dry, the lake would become a meadow before developing into a forest. Many wetland plant communities, however, are adapted to stressful conditions and have developed in concert with natural disturbances that "reset" succession or override short-term changes. So, in broad terms, succession is change in the communities of plant species. It is a continuous process that produces an ever-changing mix of species rather than a linear inevitable progression that stops with a climax community. The concept of pulse stability, where periodic floods introduce large volumes of water and nutrients to the wetland, integrates disturbance into the wetland's natural dynamics. As a result, wetland species have adapted to and, in some cases, become dependent on disturbance. Consequently, human interference with these disturbance regimes can irreversibly alter a wetland.

Natural Disturbances

Natural wetland disturbances include seasonal flooding, tidal inundation, waves, hurricanes, fire, drought, herbivory, ice scour, erosion, sedimentation, and beaver activity. Riverine floodplains, such as the vast Pantanal region in South America and the upper Nile swamps of Eastern Africa, flood during the wet season. Coastal salt marshes and mangroves are adapted to regular inundation and wave action, as well as sedimentation. Hurricanes periodically topple mangroves, which also may fall because of small localized burns ignited by lightning. Both phenomena generate the gaps that are thought to be necessary for seedling establishment. Wetlands adapted to wildfires include the Everglades, the marshes of the St. Lawrence River, and various peatlands. The prairie potholes of the Upper Midwest in the United States are adapted to drought as well as too herbivory by muskrats. Erosion and sedimentation are complementary processes in active floodplains, where water-flow shifts and a mosaic of vegetation patches develops. Beavers alter wooded landscapes by cutting trees for dam material and by impounding stream water.

Marshland in the Pantanal, south-central Brazil.

Disturbances Caused by Humans

Anthropogenic, or human-caused, disturbances include draining, diking, dredging, and filling; dam construction; logging; mining; fire suppression; and climate change. People have added water to and drained water from wetlands for millennia, and these changes have caused significant wetland loss. Several wet areas, such as the prairie potholes and the extensive freshwater wetlands of the U.S. Midwest, have been drained for agriculture. Floodplain wetlands and tidal marshes have been diked and ditched to create pastures and cropland. Dam construction has radically altered river basins by stabilizing flows, and many wetlands have been filled for building and road construction. In addition, logging has removed many of the dominant trees from various wetland types, such as bald cypress (Taxodium distichum) in the cypress swamps of the southeastern United States, and peat mining has removed centuries of accumulated organic substrate from parts of the United States, Canada, Europe, Russia, southern South America, and New Zealand.

Plant communities have responded for thousands of years to changes in climate, but future shifts may occur at an accelerated rate that results in unforeseen plant migrations, interactions, invasions, and declines. Although models designed to capture the dynamics of climate change are not precise enough to predict what will happen to a specific wetland, some broad changes are expected. On a global scale, wetlands are considered to be climate-change neutral; peatlands release methane, but they also store 15–30 percent of the world's soil carbon. If peatlands begin to decompose under warming climatic conditions, however, two greenhouse gases (carbon dioxide and methane) will be released, possibly contributing to further warming. Wetlands fed by surface water from lakes and rivers would diminish in climates that become drier, but they might expand under conditions that become wetter unless limited in geographic extent by topography and exposure to wind and waves. River systems fed by snowmelt would be particularly affected by drying conditions, as the pulse of melt water during the growing season would decrease or disappear. Wetlands fed by groundwater would expand in a wetter climate and diminish in a drier climate, though more slowly than other wetlands. Although the prairie potholes of North America are adapted to drought cycles, these wetlands could dry up completely as the climate warms, and the migratory waterfowl that use them as breeding grounds would decline. A warming climate will have hard-to-predict effects in northern latitudes, where seasonal melting of frozen ground controls the existence of wetlands in the landscape. As sea level rises with temperature, tidally influenced coastal systems will experience increased inundation. Some of these wetlands may migrate inland in areas where human infrastructure and topography (e.g., steep hillsides) do not create barriers.

Wetland Management

Wetlands can be easily altered or destroyed. They can be isolated from their water source if drainage areas are modified or impoundments are built. Major cities in the United States, such as Chicago and Washington, D.C., are located on sites that were, in part, covered by wetlands. As these cities grew, however, most of the wetlands were either drained and filled or otherwise altered substantially. The amount of wetlands lost worldwide is almost impossible to determine. It is known, though, that in the lower 48 U.S. states, a relatively newly developed region of the world, more than half of the original wetlands have been lost, primarily through conversion to agricultural land.

Humans have utilized wetlands for centuries, and countless plant and animal products are harvested from wetlands worldwide. Ancient civilizations—such as those of Mesopotamia and Egypt as well as, among the pre-Columbian civilizations, the Aztecs—developed unique systems of water delivery that involved wetlands. For centuries, salt marshes in northern Europe and the British Isles, and later in New England, were used to graze animals and raise crops of hay. Thatched roofs and fences were built from materials retrieved from wetlands. Techniques to produce fish within rice paddies or shallow ponds were developed several thousand years ago in China and Southeast Asia; crayfish harvesting is still practiced in the wetlands of Louisiana and the Philippines. In the United States, a thriving modern industry continues to depend on the harvest of cranberries from bogs. The Russians and the Irish, among others, have mined their peatlands for several centuries as a source of energy, and many countries throughout South and Southeast Asia, East Africa, and Central and South America continue to depend on mangrove wetlands for timber, food, and tannins (such as those that occur in teas and wines).

Recognition of the importance of wetlands has grown, and thus many wetlands have been protected from development by local and national policies, as well as by international projects. Examples of these efforts include the Ramsar Convention, an international agreement designed to protect the habitat of migratory waterfowl and other avian life, and the North American Waterfowl Management Plan, which was created to achieve similar goals. Wetland recognition and protection has become one of the most important facets of global natural resource protection. In addition, effective wetland management, preservation, and restoration will continue to be an important component of plans designed to mitigate the effects of global warming and climate change.

Terrestrial Ecosystem

A terrestrial ecosystem is an ecosystem found only on landforms. Six primary terrestrial ecosystems exist: tundra, taiga, temperate deciduous forest, tropical rain forest, grassland and desert.

Taiga

Taigas are cold-climate forests found in the northern latitudes.

Taigas are the world's largest terrestrial ecosystem and account for about 29% of the Earth's forests. The largest taiga ecosystems are found in Canada and Russia.

Taigas are known for their sub-arctic climate with extremely cold winters and mild summers.

They primarily consist of coniferous trees, such as pines, although there are some other deciduous trees, such as spruce and elm that have adapted to live in these areas that receive little direct sunlight for much of the year.

Taigas are home to large herbivores, such as moose, elk, and bison, as well as omnivores, such as bears.

Tundra

The tundra ecosystems of the world are found primarily north of the Arctic Circle.

They consist of short vegetation and essentially no trees.

The soil is frozen and covered with permafrost for a large portion of the year.

Caribou, polar bears, and musk ox are some of the notable species who call the tundra home.

Temperate

Temperate forests are the regions which have seasonal variation in climate i.e., the climate changes a lot from summer to winter.

The annual rain fall is about 750- 2000 mm and soil is rich. Such types of forests are found in western and central Europe, Eastern Asia and eastern North America.

These forests have deciduous trees (oaks, maples etc.) and coniferous trees (pines).

These forests contain abundant micro-organ- isms, mammals (hares, deer, fares, coyotesetc). Birds (warblers, wood peckers, owls etc.) snakes, frogs, salamanders etc.

Tropical Rain forests

Tropical rain forests are special ecosystems which accommodate thousands of species of animals and plants.

These are usually densely packed tall trees those form a ceiling from the sun above. The filing prevents the growth of smaller plants.

The temperature remains almost same throughout the year.

Such types of forests are found in Brazil of South America (Neotropic) and Central and West Africa. The area is always warm and muggy

Grasslands

Grasslands are areas dominated by grasses. They occupy about 20% of the land on the earth surface.

Grasslands occur in both in tropical and temperate regions where rainfall is not enough to support the growth of trees.

Grasslands are found in areas having well defined hot and dry, warm and rainy seasons.

Place	Name of the Grassland
North America	Prairies
Eurasia (Europe and Asia)	Steppes
Africa	Savanna
South America	Pampas
India	Grassland, Savanna

Deserts

Desert is hot and low rain areas suffering from water shortage and high wind velocity.

They show extremes of temperature. Globally deserts occupy about 1/7th of the earth's surface.

Desert animals include shrew, fox, wood rats, rabbits, camels and goat are common mammals in desert.

Other prominent desert animals are, reptiles, and burrowing rodents insects.

Man Made Changes

All natural environments and ecosystems now have an unprecedented problem to deal with humanity. Humans have brought about profound changes in a few centuries which would otherwise be expected over thousands or millions of years. The full impact of these remains to be accurately estimated. Major human impacts on ecosystems include the following:

Habitat Destruction and Fragmentation

The most direct impact of humans on ecosystems is in their destruction or conversion. Clear-cutting (the cutting of all trees within a given forest area) will, obviously, destroy a forest ecosystem. Selective logging may also alter forest ecosystems in important ways. Fragmentation- the division of a once continuous ecosystem into a number of smaller patches- may disrupt ecological processes so that the remaining areas can no longer function as they once did.

Climate Change

It is now widely accepted that human activities are contributing to global warming, chiefly through the accumulation of "greenhouse" gases in the atmosphere. The impact of this is likely to increase in the future. As noted above, climate change is a natural feature of the Earth. Previously, however, its effects were mitigated as ecosystems could effectively "migrate" by moving latitude or altitude as the climate changed. Today, so much of the world's land surface has been appropriated by people that in many cases there is no such place for the remaining natural or semi-natural ecosystems to migrate to.

Pollution

Contamination of the natural environment through a range of pollutants- herbicides, pesticides, fertilizers, industrial effluents, and human waste products- is one of the most pernicious forms of impact on the natural environment. Pollutants are often invisible, and the effects of air pollution and water pollution may not be immediately obvious, although they can be devastating in the long run.

Introduced Species

Human beings have been responsible either deliberately or accidentally for altering the distribution of a vast range of animal and plant species. This includes not only domesticated animals and cultivated plants but pests such as rats, mice, and many insects and fungi. Species which become naturalized may have a devastating impact, through predation and competition, on natural ecosystems, particularly on islands where native species have evolved in isolation. For instance, foxes, rabbits, cane toads, feral cats, and even buffaloes and camels have wreaked havoc in many ecosystems in Australia. Plants such as the South American shrub Lantana have invaded natural forests in many tropical and subtropical islands, causing major changes to these ecosystems, while the African water hyacinth Eichhornia has similarly disrupted freshwater ecosystems in many of the warmer parts of the world.

Over-Harvest

Removal of excessive numbers of animals or plants from a system can cause major ecological changes. The most important example of this at present is the over-fishing of the world's oceans. Depletion of the great majority of accessible fish stocks is undoubtedly a cause of major change, although its long-term impact is difficult to assess.

Landscape Ecology

What is Landscape Ecology?

Landscape ecology, as the name implies, is the study of landscapes; specifically, the composition, structure and function of landscapes. But what's a 'landscape'? Although there are myriad ways to define 'landscape' depending on the phenomenon under consideration, suffice it to say that a landscape is not necessarily defined by its size; rather, it is defined by an interacting mosaic of elements (e.g., ecosystems) relevant to some phenomenon under consideration (at any scale). Thus, a landscape is simply an area of land (at any scale) containing an interesting pattern that affects and is affected by an ecological process of interest. Landscape ecology, then, involves the study of these landscape patterns, the interactions among the elements of this pattern, and how these patterns and interactions change over time. In addition, landscape ecology involves the application of these principles in the formulation and solving of real-world problems.

Landscape ecology is perhaps best distinguished by its focus on:

1) Spatial heterogeneity,
2) Broader spatial extents than those traditionally studied in ecology, and,
3) The role of humans in creating and affecting landscape patterns and process.

- Spatial Heterogeneity

Landscape ecology might be defined best by its focus on spatial heterogeneity and pattern: how to characterize it, where it comes from, how it changes through time, why it matters, and how humans manage it. As such, landscape ecology has five central themes:

- Detecting pattern and the scale at which it is expressed, and summarizing it quantitatively.
- Identifying and describing the agents of pattern formation, which include the

Physical abiotic template, demographic responses to this template, and disturbance regimes overlaid on these.

- Characterizing the changes in pattern and process over space and time; that is, the dynamics of the landscape, and summarizing it quantitatively. An interest in landscape dynamics necessarily invokes models of some sort--because landscape is large and they change (usually) over time scales that are difficult to embrace empirically.
- Understanding the ecological implications of pattern; that is, why it matters to populations, communities, and ecosystems – and this is the stuff of conservation biology and ecosystem management.
- Managing landscapes to achieve human objectives.

- Broad Spatial Extents

Landscape ecology is distinguished by its focus on broader spatial extents than those traditionally studied in ecology. This stems from the anthropocentric origins of the discipline. Initial impetus for the discipline came from the geographers aerial view of the environment, for example, the patterns in the environment visible from an aerial photograph. The focus on large geographic areas is consistent with how humans typically see the world–through a coarse lens. However, modern landscape ecology does not define, a priori, specific scales that may be universally applied; rather, the emphasis is to identify scales that best characterize relationships between spatial heterogeneity and the process of interest.

- The Role of Humans

Landscape ecology is often defined by its focus on the role of humans in creating and affecting landscape patterns and process. Indeed, landscape ecology is sometimes considered to be an interdisciplinary science dealing with the interrelation between human society and its living environment. Hence, a great deal of landscape ecology deals with 'built' environments, where humans are the dominant force of landscape change. However, modern landscape ecology, with 2 its emphasis on the interplay between spatial heterogeneity and ecological process, considers humans as one of many important agents affecting landscapes, and emphasizes both natural, semi-natural, and built landscapes.

Organisms Interact with the Landscape

This question is the focus of a vast amount of research effort globally, but the following are some notable examples to give the general idea:

- Patches: species may prefer a certain kind of habitat, for example mature woodland, or ponds. Individuals of the species of concern may not be able to breed or feed outside of this type of habitat. The habitat thus defines the patch.
- Matrix: If one land use dominates the landscape, that land use forms the matrix, e.g. arable land in eastern England. If the dominant land use is uniformly inhospitable, organisms become isolated in patches of suitable habitat. For example, some characteristic plants of ancient woodland cannot survive in arable fields, and do not have seeds equipped with a mechanism to disperse between isolated woodland fragments.
- Corridors: There has been a lot of research and debate about the role of hedgerows as corridors for small woodland mammals and birds. Such species may be able to move between woodland habitat patches along hedgerows, whereas they might not feel safe enough to cross an arable field.
- Barriers: Roads, pipelines or fences might form barriers to movement of shy or less agile animals.
- Mosaic: Lesser horseshoe bats are an example of an animal that needs to live in a landscape mosaic. They sleep in old trees in mature woodland, then fly along hedgerows to open, wet places where they hunt for flies.

Landscape Ecology Important to Resource Managers

Landscape ecology (or a landscape perspective) with its focus on spatial patterns is important to resource managers because:

1) Ecosystem context matters,
2) Ecosystem function depends on the interplay of pattern and process and,
3) Because human activities can dramatically alter landscape context and the relationship between patterns and processes, resource managers have a stewardship responsibility to understand and manage these impacts – more pragmatically, resource managers have a policy and legal mandate to include a landscape perspective in resource management decisions.

- Because ecosystem context matters

Landscape ecology is founded on the principle that ecosystem composition, structure and function partially depend on the spatial (and temporal) context of the ecosystem (i.e., its landscape context); i.e., that what we observe ecologically at any particular location is affected by what is around that location. This shift in perspective from the site to the site embedded in a landscape context has profound implications for resource management. Let's consider a couple of examples:

- Metapopulations: Metapopulation depend on the number and spatial arrangement of habitat patches – where the probability of a habitat patch being occupied at any time is at least partially dependent on its proximity to other habitat patches. Focusing management on the individual site, in this case, without consideration of its landscape context, can have disastrous consequences for the population.
- Forest succession: Neighbourhood effects can play an important role in determining the successional response following a disturbance. For example, edge effects that modify the distribution of energy and water and the plant species composition of the immediate neighbourhood (which can influence the relative abundance of propagules) can exert a strong influence on succession in forest gaps and in larger openings, e.g., via wave-form succession. Ignoring these effects can lead to undesirable outcomes, including an unwanted shift in species composition or an inadequate recovery of vegetation altogether.

- Because ecosystem function depends on the interplay of pattern and process

Landscape ecology is also founded on the principle that spatial patterns affect ecological processes, which in turn affect spatial patterns. This interplay of spatial pattern and process is in fact the overarching focus of landscape ecology. While it can be argued that "ecology" has always sought to explain the relationship between pattern and process, it is safe to assert that "landscape" ecology has shifted the focus to pattern-process relationships over broad spatial extents and emphasized the role of humans in creating and affecting these relationships. This shift has profound implications for resource managers. Let's consider a couple of examples:

- Habitat fragmentation: Disruption of habitat connectivity is a major impact of human activities on plant and animal populations and one of the leading causes of the biodiversity crisis. Anthropogenic landscape elements (e.g., roads, developed land, dams) can function as impediments to the movement of organisms across the landscape, and the cumulative impacts of these impediments over broad spatial extents can be devastating.

- Alteration of disturbance regimes: Disruption of natural disturbance regimes has long lasting ecological and socioeconomic impacts. For example, disruption of fuel mass and continuity by human land use practices (e.g., livestock grazing) over broad spatial extents can dramatically alter fire regimes in fire-dependent ecosystems, leading to shifts species distributions and community structure and serious socio-economic consequences (e.g., catastrophic fires resulting in loss of life and property).

References

- Giller, S.; B. Malmqvist (1998). The Biology of Streams and Rivers. Oxford University Press, Oxford. p. 296. ISBN 0198549776
- What-is-an-ecosystem: conserve-energy-future.com, Retrieved 21 April 2018
- Energy-flow-in-an-ecosystem-with-diagram-6740: biologydiscussion.com, Retrieved 31 March 2018
- Browne, R. A. (1981). "Lakes as islands: biogeographic distribution, turnover rates, and species composition in the lakes of central New York". Journal of Biogeography. 8 1: 75–83. doi:10.2307/2844594. JSTOR 2844594
- Primary-consumer-definition-examples-and-functions: eartheclipse.com, Retrieved 19 May 2018
- What-are-secondary-consumers: worldatlas.com, Retrieved 19 July 2018
- Moss, B. (1998). Ecology of Freshwaters: man and medium, past to future. Blackwell Science, London. p. 557. ISBN 0632035129
- Tertiary-consumer-definition-examples-functions: eartheclipse.com, Retrieved 29 April 2018
- Ecosystem-services, science: britannica.com, Retrieved 25 June 2018
- Hillebrand, H. (2004). "On the generality of the latitudinal diversity gradient". American Naturalist. 163 (2): 192–211. doi:10.1086/381004. PMID 14970922

5

Ecological Economics

A transdisciplinary and interdisciplinary branch of academic research that addresses the coevolution and interdependence of natural ecosystems and human economies is called ecological economics. The diverse aspects of ecological economics such as green politics, natural capital, carrying capacity, degrowth, etc. have been thoroughly discussed in this chapter.

Ecological Economics addresses the relationships between ecosystems and economic systems in the broadest sense. These relationships are the locus of many of our most pressing current problems (i.e. sustainability, acid rain, global warming, species extinction, wealth distribution) but they are not well covered by any existing discipline. Ecological economics is a growing trans disciplinary field that aims to improve and expand economic theory to integrate the earth's natural systems, human values and human health and well-being. In conventional economics, the primary goal is to increase goods and services produced by human industries (built capital), and the gross domestic product (GDP) is a national measure of the total value of goods and services produced annually.

Ecological economics takes a broader perspective and recognizes that there are more things that contribute to human well-being than just the amount of stuff, such as health and education (human capital), friends and family (social capital) and the contribution of the earth and its biological and physical systems (natural capital). Its goal is to develop a deeper scientific understanding of the complex linkages between human and natural systems, and to use that understanding to develop effective policies that will lead to a world which is ecologically sustainable, has a fair distribution of resources (both between groups and generations of humans and between humans and other species), and efficiently allocates scarce resources including "natural" and "social" capital. Four main types of capital are considered in ecological economics:

- Built capital is the type of capital we are used to thinking about in conventional economics. It refers to goods and services created by human industry – buildings, cars, appliances, roads, toys, etc.
- Natural capital is a concept that recognizes the importance and value of the goods and services provided by nature. It goes beyond the traditional consideration of natural products as raw materials for conversion into goods (trees into houses or paper) to consider functions that are provided by planetary systems (breathable air, a stable climate) or local ecosystems (flood protection by coastal wetlands, and drinking water purification by forests). Sometimes the monetary value of these ecosystem services can be calculated, such as the cost to replace them with built capital (coastal wetlands with levees). However, some natural services are priceless. They are essential for life, and irreplaceable (breathable air).
- Social capital refers to the positive benefits gained through our interactions with others (friends, family, social groups) and the common structures of our society (languages, institutions, educational system, laws). It makes a major contribution to our collective well-being, but is hard to quantify in monetary terms.

- Human capital refers to the sum of our own health, personal experiences, education, talents, skills and interests. Collective human capital (and social capital) cannot be maximized unless there is social justice, equivalent access to the opportunities that our society provides.

Nature and Ecology

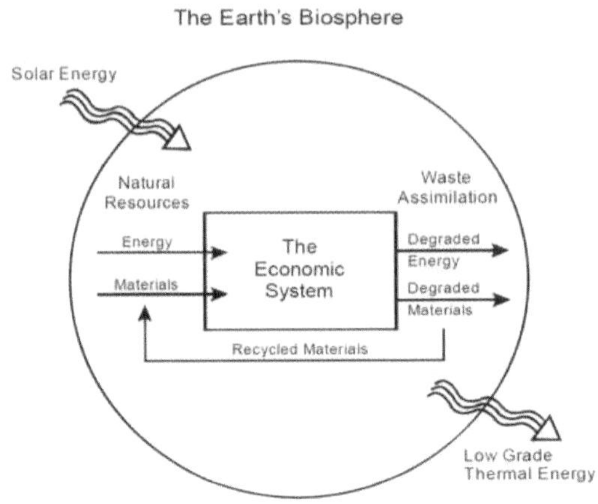

Natural resources flow through the economy and end up as waste and pollution

A simple circular flow of income diagram is replaced in ecological economics by a more complex flow diagram reflecting the input of solar energy, which sustains natural inputs and environmental services which are then used as units of production. Once consumed, natural inputs pass out of the economy as pollution and waste. The potential of an environment to provide services and materials is referred to as an "environment's source function", and this function is depleted as resources are consumed or pollution contaminates the resources. The "sink function" describes an environment's ability to absorb and render harmless waste and pollution: when waste output exceeds the limit of the sink function, long-term damage occurs. Some persistent pollutants, such as some organic pollutants and nuclear waste are absorbed very slowly or not at all; ecological economists emphasize minimizing "cumulative pollutants". Pollutants affect human health and the health of the ecosystem.

The economic value of natural capital and ecosystem services is accepted by mainstream environmental economics, but is emphasized as especially important in ecological economics. Ecological economists may begin by estimating how to maintain a stable environment before assessing the cost in dollar terms.

The Earth's carrying capacity is a central issue in ecological economics. Early economists such as Thomas Malthus pointed out the finite carrying capacity of the earth, which was also central to the MIT study *Limits to Growth*. Diminishing returns suggest that productivity increases will slow if major technological progress is not made. Food production may become a problem, as erosion, an impending water crisis, and soil salinity (from irrigation) reduce the productivity of agriculture. Ecological economists argue that industrial agriculture, which exacerbates these problems, is not sustainable agriculture, and are generally inclined favorably to organic farming, which also reduces the output of carbon.

Global wild fisheries are believed to have peaked and begun a decline, with valuable habitat such as estuaries in critical condition. The aquaculture or farming of piscivorous fish, like salmon, does not help solve the problem because they need to be fed products from other fish. Studies have shown that salmon farming has major negative impacts on wild salmon, as well as the forage fish that need to be caught to feed them.

Since animals are higher on the trophic level, they are less efficient sources of food energy. Reduced consumption of meat would reduce the demand for food, but as nations develop, they tend to adopt high-meat diets similar to that of the United States. Genetically modified food (GMF) a conventional solution to the problem, presents numerous problems – Bt corn produces its own Bacillus thuringiensis toxin/protein, but the pest resistance is believed to be only a matter of time. The overall effect of GMF on yields is contentious, with the USDA and FAO acknowledging that GMFs do not necessarily have higher yields and may even have reduced yields.

Global warming is now widely acknowledged as a major issue, with all national scientific academies expressing agreement on the importance of the issue. As the population growth intensifies and energy demand increases, the world faces an energy crisis. Some economists and scientists forecast a global ecological crisis if energy use is not contained – the Stern report is an example. The disagreement has sparked a vigorous debate on issue of discounting and intergenerational equity.

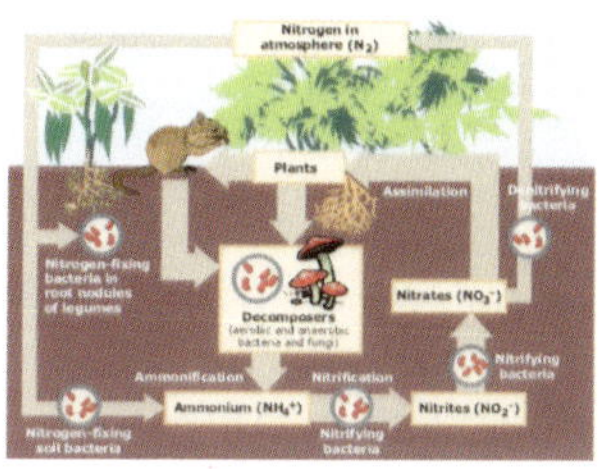

Nitrogen cycle

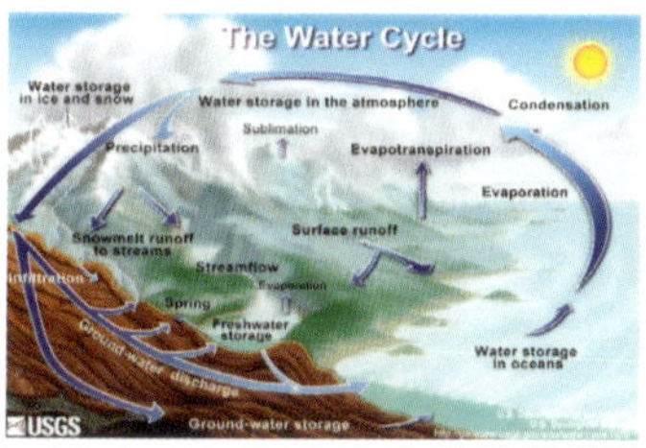

Water cycle

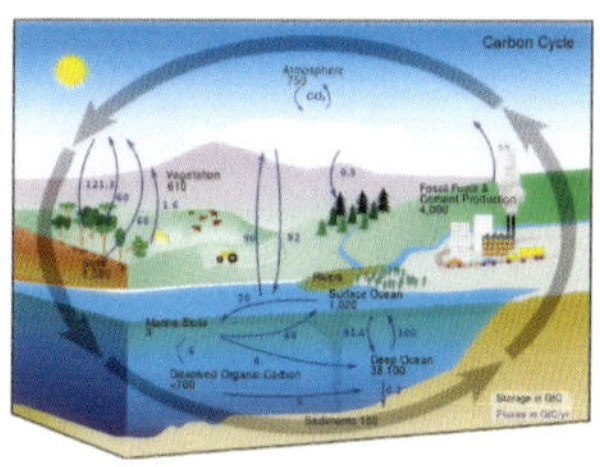

Carbon cycle

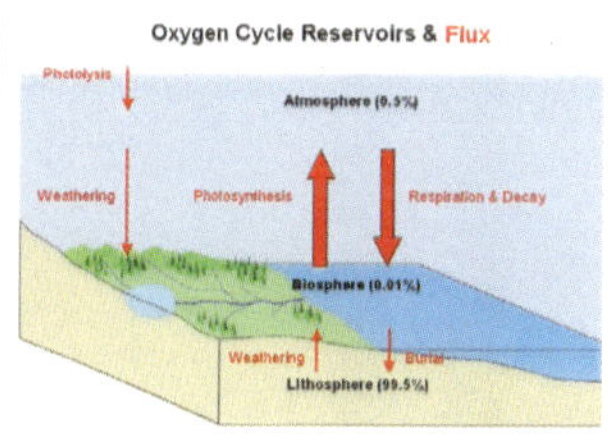

Oxygen cycle

Ethics

Mainstream economics has attempted to become a value-free 'hard science', but ecological economists argue that value-free economics is generally not realistic. Ecological economics is more willing to entertain alternative conceptions of utility, efficiency, and cost-benefits such as positional analysis or multi-criteria analysis. Ecological economics is typically viewed as economics for sustainable development, and may have goals similar to green politics.

Green Economics

In international, regional, and national policy circles, the concept of the *green economy* grew in popularity as a response to the financial predicament at first then became a vehicle for growth and development.

The *United Nations Environment Program* (UNEP) defines a 'green economy' as one that focuses on the human aspects and natural influences and an economic order that can generate high-salary jobs. In 2011, its definition was further developed as the word 'green' is made to refer to an economy that is not only resourceful and well-organized but also impartial, guaranteeing an objective shift to an economy that is low-carbon, resource-efficient, and socially-inclusive.

The ideas and studies regarding the green economy denote a fundamental shift for more effective, resourceful, environment-friendly and resource-saving technologies that could lessen emissions and alleviate the adverse consequences of climate change, at the same time confront issues about resource exhaustion and grave environmental dilapidation.

As an indispensable requirement and vital precondition to realizing sustainable development, the Green Economy adherents robustly promote good governance. To boost local investments and foreign ventures, it is crucial to have a constant and foreseeable macroeconomic atmosphere. Likewise, such an environment will also need to be transparent and accountable. In the absence of a substantial and solid governance structure, the prospect of shifting towards a sustainable development route would be insignificant. In achieving a green economy, competent institutions and governance systems are vital in guaranteeing the efficient execution of strategies, guidelines, campaigns, and programmes.

Shifting to a Green Economy demands a fresh mindset and an innovative outlook of doing business. It likewise necessitates new capacities, skills set from labor and professionals who can competently function across sectors, and able to work as effective components within multi-disciplinary teams. To achieve this goal, vocational training packages must be developed with focus on greening the sectors. Simultaneously, the educational system needs to be assessed as well in order to fit in the environmental and social considerations of various disciplines.

Green Politics

Schools of Thought

Various competing schools of thought exist in the field. Some are close to resource and environmental economics while others are far more heterodox in outlook. An example of the latter is the *European Society for Ecological Economics*. An example of the former is the Swedish *Beijer International Institute of Ecological Economics*. Clive Spash has argued for the classification of the ecological economics movement, and more generally work by different economic schools on the environment, into three main categories. These are the mainstream new resource economists, the new environmental pragmatists, and the more radical social ecological economists. International survey work comparing the relevance of the categories for mainstream and heterodox economists shows some clear divisions between environmental and ecological economists.

Differentiation from Mainstream Schools

Some ecological economists prioritise adding natural capital to the typical capital asset analysis of land, labor, and financial capital. These ecological economists then use tools from mathematical economics as in mainstream economics, but may apply them more closely to the natural world. Whereas mainstream economists tend to be technological optimists, ecological economists are inclined to be technological sceptics. They reason that the natural world has a limited carrying capacity and that its resources may run out. Since destruction of important environmental resources could be practically irreversible and catastrophic, ecological economists are inclined to justify cautionary measures based on the precautionary principle.

The most cogent example of how the different theories treat similar assets is tropical rainforest ecosystems, most obviously the Yasuni region of Ecuador. While this area has substantial deposits of bitumen it is also one of the most diverse ecosystems on Earth and some estimates establish it has over 200 undiscovered medical substances in its genomes - most of which would be destroyed by logging the forest or mining the bitumen. Effectively, the instructional capital of the genomes is undervalued by analyses which view the rainforest primarily as a source of wood, oil/tar and perhaps food. Increasingly the carbon credit for leaving the extremely carbon-intensive ("dirty") bitumen in the ground is also valued - the government of Ecuador set a price of US$350M for an oil lease with the intent of selling it to someone committed to never exercising it at all and instead preserving the rainforest.

While this natural capital and ecosystems services approach has proven popular amongst many it has also been contested as failing to address the underlying problems with mainstream economics, growth, market capitalism and monetary valuation of the environment. Critiques concern the need to create a more meaningful relationship with Nature and the non-human world than evident in the instrumentalism of shallow ecology and the environmental economists commodification of everything external to the market system.

Among the topics addressed by ecological economics is methodology, allocation of resources, weak versus strong sustainability, energy economics, energy accounting and balance, environmental services, cost shifting, and modeling.

Methodology

A primary objective of ecological economics (EE) is to ground economic thinking and practice in physical reality, especially in the laws of physics (particularly the laws of thermodynamics) and in knowledge of biological systems. It accepts as a goal the improvement of human well-being through development, and seeks to ensure achievement of this through planning for the sustainable development of ecosystems and societies. Of course the terms development and sustainable development are far from lacking controversy. Richard B. Norgaard argues traditional economics has hi-jacked the development terminology in his book *Development Betrayed.*

Well-being in ecological economics is also differentiated from welfare as found in mainstream economics and the 'new welfare economics' from the 1930s which informs resource and environmental economics. This entails a limited preference utilitarian conception of value i.e., Nature is valuable to our economies, that is because people will pay for its services such as clean air, clean water, encounters with wilderness, etc.

Ecological economics is distinguishable from neoclassical economics primarily by its assertion that the economy is embedded within an environmental system. Ecology deals with the energy and matter transactions of life and the Earth, and the human economy is by definition contained within this system. Ecological economists argue that neoclassical economics has ignored the environment, at best considering it to be a subset of the human economy.

The neoclassical view ignores much of what the natural sciences have taught us about the contributions of nature to the creation of wealth e.g., the planetary endowment of scarce matter and energy, along with the complex and biologically diverse ecosystems that provide goods and ecosystem services directly to human communities: micro- and macro-climate regulation, water recycling, water purification, storm water regulation, waste absorption, food and medicine production, pollination, protection from solar and cosmic radiation, the view of a starry night sky, etc.

There has then been a move to regard such things as natural capital and ecosystems functions as goods and services. However, this is far from uncontroversial within ecology or ecological economics due to the potential for narrowing down values to those found in mainstream economics and the danger of merely regarding Nature as a commodity. This has been referred to as ecologists 'selling out on Nature'. There is then a concern that ecological economics has failed to learn from the extensive literature in environmental ethics about how to structure a plural value system.

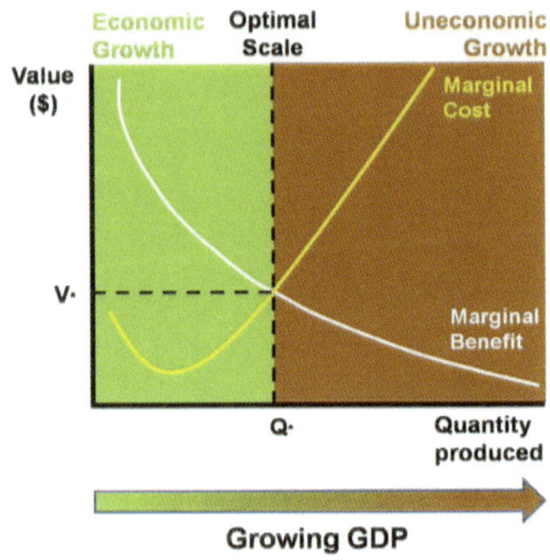

The marginal costs of a growing economy may gradually exceed the marginal benefits, however measured.

Allocation of Resources

Resource and neoclassical economics focus primarily on the efficient allocation of resources and less on the two other problems of importance to ecological economics: distribution (equity), and the scale of the economy relative to the ecosystems upon which it relies. Ecological economics makes a clear distinction between growth (quantitative increase in economic output) and development (qualitative improvement of the quality of life), while arguing that neoclassical economics confuses the two. Ecological economists point out that beyond modest levels, increased per-capita consumption (the typical economic measure of "standard of living") may not always lead to improvement in human well-being, but may have harmful effects on the environment and broader societal well-being. This situation is sometimes referred to as uneconomic growth.

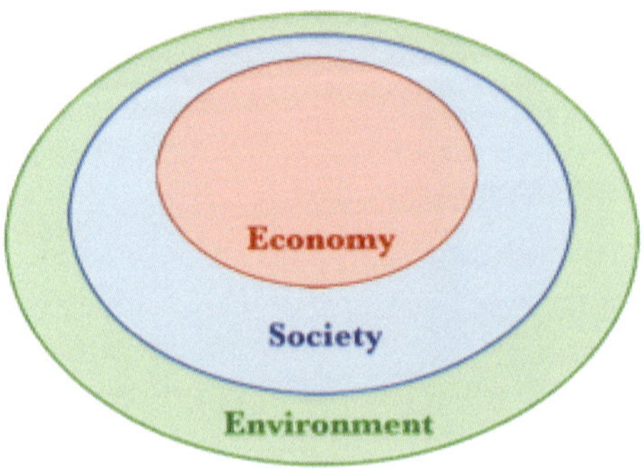

The three nested systems of sustainability - the economy wholly contained by society, wholly contained by the biophysical environment. Clickable.

Weak versus Strong Sustainability

Ecological economics challenges the conventional approach towards natural resources, claiming that it undervalues natural capital by considering it as interchangeable with human-made capital—labor and technology.

The impending depletion of natural resources and increase of climate-changing greenhouse gasses should motivate us to examine how political, economic and social policies can benefit from alternative energy. Shifting dependence on fossil fuels with specific interest within just one of the above-mentioned factors easily benefits at least one other. For instance, photo voltaic (or solar) panels have a 15% efficiency when absorbing the sun's energy, but its construction demand has increased 120% within both commercial and residential properties. Additionally, this construction has led to a roughly 30% increase in work demands.

The potential for the substitution of man-made capital for natural capital is an important debate in ecological economics and the economics of sustainability. There is a continuum of views among economists between the strongly neoclassical positions of Robert Solow and Martin Weitzman, at one extreme and the 'entropy pessimists', notably Nicholas Georgescu-Roegen and Herman Daly, at the other.

Neoclassical economists tend to maintain that man-made capital can, in principle, replace all types of natural capital. This is known as the weak sustainability view, essentially that every technology can be improved upon or replaced by innovation, and that there is a substitute for any and all scarce materials.

At the other extreme, the strong sustainability view argues that the stock of natural resources and ecological functions are irreplaceable. From the premises of strong sustainability, it follows that economic policy has a fiduciary responsibility to the greater ecological world, and that sustainable development must therefore take a different approach to valuing natural resources and ecological functions.

Recently, Stanislav Shmelev developed a new methodology for the assessment of progress at the macro scale based on multi-criteria methods, which allows consideration of different perspectives, including strong and weak sustainability or conservationists vs industrialists and aims to search for a 'middle way' by providing a strong neo-Keynesian economic push without putting excessive pressure on the natural resources, including water or producing emissions, both directly and indirectly.

Energy Economics

A key concept of energy economics is net energy gain, which recognizes that all energy requires energy to produce. To be useful the energy return on energy invested (*EROEI*) has to be greater than one. The net energy gain from production coal, oil and gas has declined over time as the easiest to produce sources have been most heavily depleted.

Ecological economics generally rejects the view of energy economics that growth in the energy supply is related directly to well being, focusing instead on biodiversity and creativity - or natural capital and individual capital, in the terminology sometimes adopted to describe these economically. In practice, ecological economics focuses primarily on the key issues of uneconomic growth and quality of life. Ecological economists are inclined to acknowledge that much of what is important in human well-being is not analyzable from a strictly economic standpoint and suggests an interdisciplinary approach combining social and natural sciences as a means to address this.

Ecological economics generally rejects the view of energy economics that growth in the energy supply is related directly to well being, focusing instead on biodiversity and creativity - or natural capital and individual capital, in the terminology sometimes adopted to describe these economically. In practice, ecological economics focuses primarily on the key issues of uneconomic growth and quality of life. Ecological economists are inclined to acknowledge that much of what is important in human well-being is not analyzable from a strictly economic standpoint and suggests an interdisciplinary approach combining social and natural sciences as a means to address this.

Thermoeconomics is based on the proposition that the role of energy in biological evolution should be defined and understood through the second law of thermodynamics, but also in terms of such economic criteria as productivity, efficiency, and especially the costs and benefits (or profitability) of the various mechanisms for capturing and utilizing available energy to build biomass and do work. As a result, thermoeconomics is often discussed in the field of ecological economics, which itself is related to the fields of sustainability and sustainable development.

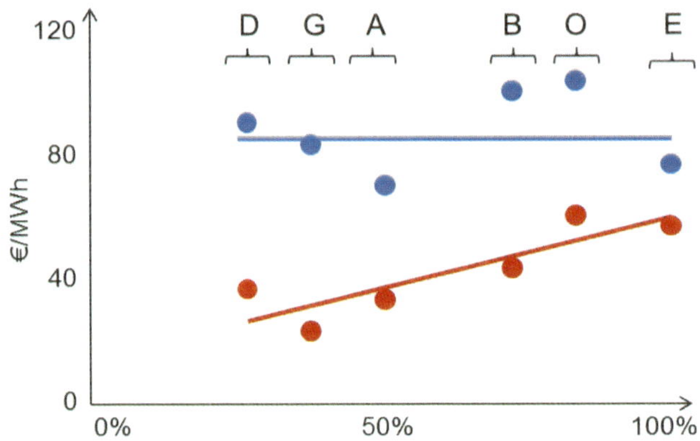

Exergy analysis can be performed to find connections between economic value and the physical world.

Here the costs of heating (vertical axis) are compared with the exergy content of different energy carriers (horizontal axis). Red dots and trend line indicate energy prices for consumers, blue dots and trend line indicate total price for consumers including capital expenditure for the heating system. Energy carriers included are district heating (D), ground-source heat pump (G), exhaust air heat pump (A), bioenergy meaning firewood (B), heating oil (O) and direct electric heating (E).

Exergy analysis is performed in the field of industrial ecology to use energy more efficiently. The term *exergy*, was coined by Zoran Rant in 1956, but the concept was developed by J. Willard Gibbs. In recent decades, utilization of exergy has spread outside of physics and engineering to the fields of industrial ecology, ecological economics, systems ecology, and energetics.

Energy Accounting and Balance

An energy balance can be used to track energy through a system, and is a very useful tool for determining resource use and environmental impacts, using the First and Second laws of thermodynamics, to determine how much energy is needed at each point in a system, and in what form that energy is a cost in various environmental issues. The energy accounting system keeps track of energy in, energy out, and non-useful energy versus work done, and transformations within the system.

Scientists have written and speculated on different aspects of energy accounting.

Ecosystem Services and their Valuation

Ecological economists agree that ecosystems produce enormous flows of goods and services to human beings, playing a key role in producing well-being. At the same time, there is intense debate about how and when to place values on these benefits.

A study was carried out by Costanza and colleagues to determine the 'value' of the services provided by the environment. This was determined by averaging values obtained from a range of studies conducted in very specific context and then transferring these without regard to that context. Dollar figures were averaged to a per hectare number for different types of ecosystem e.g. wetlands, oceans. A total was then produced which came out at 33 trillion US dollars (1997 values), more than twice the total GDP of the world at the time of the study. This study was criticized by pre-ecological and even some environmental economists - for being inconsistent with assumptions of financial capital valuation - and ecological economists - for being inconsistent with an ecological economics focus on biological and physical indicators.

The whole idea of treating ecosystems as goods and services to be valued in monetary terms remains controversial. A common objection is that life is precious or priceless, but this demonstrably degrades to it being worthless within cost-benefit analysis and other standard economic methods. Reducing human bodies to financial values is a necessary part of mainstream economics and not always in the direct terms of insurance or wages. Economics, in principle, assumes that conflict is reduced by agreeing on voluntary contractual relations and prices instead of simply fighting or coercing or tricking others into providing goods or services. In doing so, a provider agrees to surrender time and take bodily risks and other (reputation, financial) risks. Ecosystems are no different from other bodies economically except insofar as they are far less replaceable than typical labour or commodities.

Despite these issues, many ecologists and conservation biologists are pursuing ecosystem valuation. Biodiversity measures in particular appear to be the most promising way to reconcile financial and ecological values, and there are many active efforts in this regard. The growing field of biodiversity finance began to emerge in 2008 in response to many specific proposals such as the Ecuadoran Yasuni proposal or similar ones in the Congo. US news outlets treated the stories as a "threat" to "drill a park" reflecting a previously dominant views that NGOs and governments had the primary responsibility to protect ecosystems. However Peter Barnes and other commentators have recently argued that a guardianship/trustee/commons model is far more effective and takes the decisions out of the political realm.

Commodification of other ecological relations as in carbon credit and direct payments to farmers to preserve ecosystem services are likewise examples that enable private parties to play more direct roles protecting biodiversity, but is also controversial in ecological economics. The United Nations Food and Agriculture Organization achieved near-universal agreement in 2008 that such payments directly valuing ecosystem preservation and encouraging permaculture were the only practical way out of a food crisis. The holdouts were all English-speaking countries that export GMOs and promote "free trade" agreements that facilitate their own control of the world transport network: The US, UK, Canada and Australia.

Not 'Externalities', but Cost Shifting

Ecological economics is founded upon the view that the neoclassical economics (NCE) assumption that environmental and community costs and benefits are mutually canceling *"externalities"* is not warranted. Joan Martinez Alier, for instance shows that the bulk of consumers are automatically excluded from having an impact upon the prices of commodities, as these consumers are future generations who have not been born yet. The assumptions behind future discounting, which assume that future goods will be cheaper than present goods, has been criticized by David Pearce and by the recent Stern Report (although the Stern report itself does employ discounting and has been criticized for this and other reasons by ecological economists such as Clive Spash).

Concerning these externalities, some like the eco-businessman Paul Hawken argue an orthodox economic line that the only reason why goods produced unsustainably are usually cheaper than goods produced sustainably is due to a hidden subsidy, paid by the non-monetized human environment, community or future generations. These arguments are developed further by Hawken, Amory and Hunter Lovins to promote their vision of an environmental capitalist utopia in *Natural Capitalism: Creating the Next Industrial Revolution*.

In contrast, ecological economists, like Joan Martinez-Alier, appeal to a different line of reasoning. Rather than assuming some (new) form of capitalism is the best way forward, an older ecological economic critique questions the very idea of internalizing externalities as providing some corrective to the current system. The work by Karl William Kapp explains why the concept of "externality" is a misnomer. In fact the modern business enterprise operates on the basis of shifting costs onto others as normal practice to make profits. Charles Eisenstein has argued that this method of privatising profits while socialising the costs through externalities, passing the costs to the community, to the natural environment or to future generations is inherently destructive. As social ecological economist Clive Spash has noted, externality theory fallaciously assumes environmental and social problems are minor aberrations in an otherwise perfectly functioning efficient economic system. Internalizing the odd externality does nothing to address the structural systemic problem and fails to recognize the all pervasive nature of these supposed 'externalities'.

Ecological-Economic Modeling

Mathematical modeling is a powerful tool that is used in ecological economic analysis. Various approaches and techniques include: evolutionary, input-output, neo-Austrian modeling, entropy and thermodynamic models, multi-criteria, and agent-based modeling, the environmental Kuznets curve, and Stock-Flow consistent model frameworks. System dynamics and GIS are techniques applied, among other, to spatial dynamic landscape simulation modeling. The Matrix accounting methods of Christian Felber provide a more sophisticated method for identifying "the common good".

Criticism

Assigning monetary value to natural resources such as biodiversity, and the emergent ecosystem services is often viewed as a key process in influencing economic practices, policy, and decision-making. While this idea is becoming more and more accepted among ecologists and conservationist, some argue that it is inherently false.

McCauley argues that ecological economics and the resulting ecosystem service based conservation can be harmful. He describes four main problems with this approach:

Firstly, it seems to be assumed that all ecosystem services are financially beneficial. This is undermined by a basic characteristic of ecosystems: they do not act specifically in favour of any single species. While certain services might be very useful to us, such as coastal protection from hurricanes by mangroves for example, others might cause financial or personal harm, such as wolves hunting cattle. The complexity of Eco-systems makes it challenging to weigh up the value of a given species. Wolves play a critical role in regulating prey populations; the absence of such an apex predator in the Scottish Highlands has caused the over population of deer, preventing afforestation, which increases the risk of flooding and damage to property.

Secondly, allocating monetary value to nature would make its conservation reliant on markets that fluctuate. This can lead to devaluation of services that were previously considered financially beneficial. Such is the case of the bees in a forest near former coffee plantations in Finca Santa Fe, Costa Rica. The pollination services were valued to over US$60,000 a year, but soon after the study, coffee prices dropped and the fields were replanted with pineapple. Pineapple does not require bees to be pollinated, so the value of their service dropped to zero.

Thirdly, conservation programmes for the sake of financial benefit underestimate human ingenuity to invent and replace ecosystem services by artificial means. McCauley argues that such proposals are deemed to have a short lifespan as the history of technology is about how Humanity developed artificial alternatives to nature's services and with time passing the cost of such services tend to decrease. This would also lead to the devaluation of ecosystem services.

Lastly, it should not be assumed that conserving ecosystems is always financially beneficial as opposed to alteration. In the case of the introduction of the Nile perch to Lake Victoria, the ecological consequence was decimation of native fauna. However, this same event is praised by the local communities as they gain significant financial benefits from trading the fish.

McCauley argues that, for these reasons, trying to convince decision-makers to conserve nature for monetary reasons is not the path to be followed, and instead appealing to morality is the ultimate way to campaign for the protection of nature.

Green Politics

Green Politics is a relatively recent political movement that places a concern for nature and its myriad species at the top of its agenda. New social and political movements arise in response to crises that are perceived to be both long-term and systemic. The crisis out of which the broadly based Green movement has emerged is the environmental crisis, which is actually a series of interconnected crises caused by population growth, air and water pollution, the destruction of the tropical and temperate rain forests, the rapid extinction of entire species of plants and animals, the greenhouse effect, acid rain , ozone layer depletion , and other now familiar instances of environmental degradation . Many are by-products of technological innovations, such as the internal combustion engine. But the causes of these environmental crises are not only technological but are also broadly cultural and political. They stem from beliefs and attitudes that place human beings above or apart from nature. Despite their differences, the major mainstream political perspectives–liberalism, socialism, and conservatism—are alike in viewing nature as either a hostile force to be conquered or a resource base to be exploited for human purposes. All, in short, share an anthropocentric, or human-cantered, bias.

Against these views, the modern environmental or Green movement counterpoises its own perspective. Many Greens prefer not to call their perspective a political ideology, but an environmental ethic. Earlier ecological thinkers, such as Aldo Leopold, spoke of a "land ethic." Others, such as Christopher stone, speak of an ethic with the earth itself at its centre, while still others, such as Hans Jonas, speak of an emerging "planetary ethic." Despite differences of accent and emphasis, however, all are alike in several important respects. An environmental or Ecological ethics, they say, would include several key features. First, such an ethic would emphasize the web of interconnections and mutual dependence within which we and other species live. From the recognition of interconnectedness a second feature follows: a respect for all life, however humble humans may believe it to be, because the fate of our species is tied in with theirs. And since life requires certain conditions to sustain it, the third feature follows: we have an obligation to respect and care for the conditions that nurture and sustain life in its myriad forms. Since nature nourishes her creatures within a complex web of interconnected conditions, to damage one part of this life-sustaining web is to damage the others, and to endanger the existence of the creatures that depend upon it.

Green thinkers hold that the enormous power that humans have over nature imposes on our species a special responsibility for restraining our reach and using our power wisely and well. Greens point out that the fate of the earth and all its creatures now depends, to an unprecedented degree, on human decisions and actions. For not only do we depend on nature, but nature depends on our care and restraint and forbearance. Humans have the nuclear means to destroy in mere minutes the earth's inhabitants and the ecosystems that sustain them. From this emerges a fourth feature of a "green" political perspective: Greens must oppose militarism and work for peace.

But the earth is in danger not only from global thermo-nuclear war but from the slower destruction of the natural environment. Such destruction is a consequence not only of large-scale policies but of small-scale, everyday acts. And each of us bears full responsibility for our actions and also, since we live in a democracy, some share of responsibility for cumulative effects and collective outcomes. Each of us has, or can have, a hand in making the laws under which we live. It is for this reason that Greens give equal emphasis to our collective and individual responsibility for protecting the environment that protects us. The fifth feature of the Green political perspective, then, is to emphasize the importance of informed and active democratic citizenship at the grass-roots level. Hence the Green adage, "Think globally and act locally".

On this much most Greens agree. But there are also a number of unresolved differences of approach, emphasis, and political strategy. The internal ideological spectrum ranges from "light green" conservationists to "dark green" radicals, and includes assorted anarchist beliefs, deep ecology, ecofeminism, social ecology, bioregionalism, New Age Gaia worship, and other groupings, each differing in various ways form the others. Among these are differences regarding the basic beliefs underlying and motivating the green movement. Some New Age Greens envision an environmental ethic grounded in spiritual or religious values. We should, they say, look upon the earth as a benevolent and kindly deity—the goddess Gaia (from the Greek word for "earth")—to be worshiped in reverence and awe. In this way we can liberate ourselves from the restrictive rationalism that characterizes modern science. Other Greens disapprove of such a spiritual or religious orientation, contending that such beliefs are politically pernicious and inimical to the rational scientific thinking required to diagnose and solve environmental problems.

Other differences have to do with the political strategies and tactics to be employed by the environmental movement. Some say that Greens should take an active part in electoral politics, perhaps even following the lead of Greens in Germany and organizing a Green Party. Aware of the formidable obstacles facing minority third parties, most have favoured other strategies, such as working within existing mainstream parties (especially the Democratic Party in the United States), or hiring lobbyists to influence legislation. Still other Greens favour working outside of traditional interest group politics, believing the earth and its inhabitants hardly constitute a special interest. Others, such as social ecologists, tend to favour local, grass-roots campaigns which involve neighbours, friends, and fellow citizens in efforts to protect the environment. Some social ecologists are anarchists who see the state and its pro-business and pro-growth policies as the problem, rather than the solution, and seek its eventual replacement by a decentralized system of communes and cooperatives. Greens of the "bioregionalist" persuasion add that such social and political organization ought to be based on biological or natural, rather than artificial or political, boundaries and regions.

Although all Greens agree on the importance of informing and educating the public, they disagree as to how this might best be done. Some groups, such as Greenpeace, favour dramatic direct action calculated to make headlines and capture public attention. Even more militant groups, such as the Sea Shepherd Society and Earth First! , have advocated monkey-wrenching as a morally justifiable means of publicizing and protesting practices destructive of the natural environment.

Such militant tactics are decried by moderate or mainstream groups, which tend to favour subtle, low-key efforts to influence legislation and inform the public on environmental matters. The Sierra Club, for example, lobbies Congress and state legislatures to pass environmental legislation. It also publishes books and produces films and videos about a wide variety of environmental issues. Similar strategies are followed by other groups, such as the Environmental Defence Fund. Another group, The Nature Conservancy, solicits funds to buy land for nature preserves.

Differences over strategy and tactics are, however, differences about means and not necessarily about basic assumptions and ends. Despite their political differences, Greens are alike in assuming that all things are connected—ecology is, after all, the study of interconnections—and they agree that complex ecosystem and the myriad life-forms they sustain are valuable and worthy of protection by political and other means.

Green Economics

Green economics is the economics of the real world—the world of work, human needs, the Earth's materials, and how they mesh together most harmoniously. It is primarily about "use-value", not "exchange-value" or money. It is about quality, not quantity for the sake of it. It is about regeneration---of individuals, communities and ecosystems---not about accumulation, of either money or material.

The industrial or capitalist definition of wealth has always been about the accumulation of money and matter. Any use-values generated (i.e. social needs met) have been secondary—a side-effect, by-product, spin-off or trickle-down—to the primary goal of monetary accumulation. For two centuries, the quest to accumulate money or capital drove a powerful industrialization process that actually did spin off many human benefits, however unfairly distributed. But blind material and monetary growth has reached a threshold where it is generating more destruction than real wealth. A post-industrial world requires an economics of quality, where both money and matter are returned to a status of means to an end. Green economics means a direct focus on meeting human and environmental need.

Tinkering with money, interest rates, or even state regulation is insufficient in creating sensible economies. One can scarcely imagine a more inefficient, irrational and wasteful way to organize any sector of the economy than what we actually have right now. Both the form and the content of sustainable agriculture, of green manufacturing, of soft energy, etc. are diametrically opposed to their current industrial counterparts, which are intrinsically wasteful. There is no justifiable rationale to be producing vast quantities of toxic materials; or generating more deskilled than skilled labour; or displacing labour rather than resources from production; or extending giant wasteful loops of production & consumption through globalization. These are economic inefficiencies, economic irrationalities that can only be righted by starting from scratch—to look at the most elegant and efficient ways of doing everything. As green economist Paul Hawken writes, our social and environmental crises are not problems of management, but of design. We need a system overhaul.

Green economics is not just about the environment. Certainly we must move to harmonize with natural systems, to make our economies flow benignly like sailboats in the wind of ecosystem processes. But doing this requires great human creativity, tremendous knowledge, and the widespread participation of everyone. Human beings and human workers can no longer serve as cogs in the machine of accumulation, be it capitalistic or socialistic. Ecological development requires an unleashing of human development and an extension of democracy. Social and ecological transformations go hand-in-hand.

Green economics and green politics both emphasize the creation of positive alternatives in all areas of life and every sector of the economy. Green economics does not prioritize support for either the "public" or the "private" sector. It argues that both sectors must be transformed so that markets express social and ecological values, and the state becomes merged with grassroots networks of community innovation. For this to happen, new economic processes must be designed and new rules of the game written, so that incentives for ecological conduct are built into everyday economic life. The state can then function less as a policeman, and more as a coordinator. This is a very different kind of "self-regulation" than current profit- and power- driven market forces. The basis for self-regulation in a green economy would be community, and intelligent design which provides incentives for the right things.

Principles of Green Economy

1. The Primacy of Use-value, Intrinsic Value & Quality: This is the fundamental principle of the green economy as a service economy, focused on end-use, or human and environment needs. Matter is a means to the end of satisfying real need, and can be radically conserved. Money similarly must be returned to a status as a means to facilitate regenerative exchanges, rather than an end in itself. When this is done in even a significant portion of the economy, it can undercut the totalitarian power of money in the entire economy.
2. Following Natural Flows: The economy moves like a proverbial sailboat in the wind of natural processes by flowing not only with solar, renewable and "negawatt" energy, but also with natural hydrological cycles, with regional vegetation and food webs, and with local materials. As society becomes more ecological, political and economic boundaries tend to coincide with ecosystem boundaries. That is, it becomes bioregional.
3. Waste Equals Food: In nature there is no waste, as every process output is an input for some other process. This principle implies not only a high degree of organizational complementarity, but also that outputs and by-products are nutritious and non-toxic enough to be food for something.
4. Elegance and Multifunctionality: Complex food webs are implied by the previous principle--integrated relationships which are antithetical to industrial society's segmentation and fragmentation. What Roberts & Brandum (1995) call "economics with peripheral vision", this elegance features "problem-solving strategies that develop multiple wins and positive side-effects from any one set of actions".
5. Appropriate Scale/Linked Scale: This does not simply mean "small is beautiful", but that every regenerative activity has its most appropriate scale of operation. Even the smallest activities have larger impacts, however, and truly ecological activity "integrates design across multiple scales", reflecting influence of larger on smaller and smaller on larger (Van der Ryn and Cowan, 1996).
6. Diversity: In a world of constant flux, health and stability seem to depend on diversity. This applies to all levels (diversity of species, of ecosystems, of regions), and to social as well as ecological organization.
7. Self-Reliance, Self-Organization, Self-Design: Complex systems necessarily rely on "nested hierarchies" of intelligence which coordinate among themselves in a kind of resonant dance. These hierarchies are built from the bottom up, and - in contrast to civilization's social hierarchies - the base levels are the most important. In an economy which moves with ecosystem processes, tremendous scope for local response, design and adaptation must be provided - although these local and regional domains must be attuned to larger processes. Self-reliance is not self-sufficiency, but facilitates a more flexible and holistic interdependence.
8. Participation & Direct Democracy: To enable flexibility and resilience, ecological economic design features a high "eyes to acres" ratio (Van der Ryn & Cowan, 1996) - that is, lots of

local observation and participation. Conversely, ecological organization and new information/communications technologies can provide the means for deeper levels of participation in the decisions that count in society.

9. Human Creativity and Development: Displacing resources from production and tuning into the spontaneous productivity of nature requires tremendous creativity. It requires all-round human development that entails great qualities of nurture. These are qualities of giving and real service that have been suppressed (especially in men) by the social and psychological conditioning of the industrial order. In green change, the personal and political, the social and ecological go hand-in-hand. Social, aesthetic and spiritual capacities become central to attaining economic efficiency, and become important goals in themselves.

10. The Strategic role of the Built-environment, the Landscape & Spatial Design: As Permaculturalist Bill Mollison has emphasized, the greatest efficiency gains can often be achieved by a simple spatial rearrangement of system components. Elegant, mixed-use integrated design which moves with nature is place-based. In addition, our buildings, in one way or another, absorb around 40 per cent of materials and energy throughput in North America. Thus, conservation and efficiency improvements in this sector impact tremendously on the entire economy.

Green economic conversion must be radical, but it must also be incremental and organic. How is this possible? Rodale cites the need for a kind of economic succession which mimics ecological landscape change. We need "pioneer enterprises" which can thrive in today's hostile economic landscape, but also prepare the ground for more ecological and egalitarian enterprises to come. A vision of what each sector of the economy would look like in an ecological economy--based on the specifics of each place--is a starting point. This vision must be coupled with practical action in each of these sectors, gradually moving toward this vision. Enough practical activity can eventually generate the impetus for state action to level the playing field for ecological alternatives.

Natural Capital

Natural capital is the world's stocks of natural assets including soil, air, water and all living things, which combine to create an ecosystem that makes life possible. Its most obvious benefits for humans are clean air, food, water, energy, shelter, sources of medicine, and many of the materials that we use. Less visible ecosystem services include natural flood defences provided by forests; climate regulation; the pollination of crops; and billions of tonnes of carbon stored by peatlands. And while we talk of many recognised 'capitals', including financial, manufactured, social, human, and intellectual, natural capital underpins all of these by providing the base on which societies can thrive and economies can prosper.

Natural capital has become an increasingly serious issue of concern because much like running up financial debts – which can result in bankruptcy, when we draw down too much stock from our natural environment, without ensuring and encouraging its recovery – we run the risk of local, regional and eventually, global ecosystem collapse.

In short, we have increased financial capital, in a large part through the exploitation and degradation of natural and social capital. Over-exploiting natural capital can prove to be catastrophic, not just in terms of biodiversity loss, but for humans economically and socially when ecosystem productivity and resilience decline. This is becoming increasingly evident over time, particularly in regions which are prone to extreme weather events such as floods and droughts. Communities struggle to sustain themselves in stressed ecosystems, and potentially face the risk of starvation, conflict over resource scarcity, and displacement of populations.

Many studies have assessed natural capital's value in financial terms. The United Nations tell us that business activity degrades nearly 30 million acres of land every year at a cost of USD 40 billion to the global economy. In fact, the issue of degraded land has become so urgent that at the COP21 climate negotiations in 2015, the Prince of Wales, along with many others, urged delegates to consider "landscape restoration" as much of a priority as halting its degradation. Using a science based approach, it is possible to rebuild hydrology and replant native trees or plants to restore natural capital, but businesses need to be convinced, and a clear case has to be established by measuring and accounting for the value that ecosystems provide.

Natural Capital Declaration

In June 2012 a 'natural capital declaration' (NCD) was launched at the Rio+20 summit held in Brazil. An initiative of the global finance sector, it was signed by 40 CEOs to 'integrate natural capital considerations into loans, equity, fixed income and insurance products, as well as in accounting, disclosure and reporting frameworks.' They worked with supporting organisations to develop tools and metrics to integrate natural capital factors into existing business structures. In summary, its four key aims are to:

- Increase understanding of business dependency on natural capital assets;
- Support development of tools to integrate natural capital considerations into the decision-making process of all financial products and services;
- Help build a global consensus on integrating natural capital into private sector accounting and decision-making;
- Encourage a consensus on integrated reporting to include natural capital as one of the key components of an organisation's success.

Natural Capital Protocol

In July 2016, the Natural Capital Coalition released the Natural Capital Protocol. The Protocol provides a standardised framework for organisations to identify measure and value their direct and indirect impacts and dependencies on natural capital. The Protocol harmonises existing tools and methodologies, and guides organisations towards the information they need to make strategic and operational decisions that include impacts and dependencies on natural capital.

The Protocol was developed in a unique collaboration between 38 organisations who signed voluntary, pre-competitive contracts.

The Protocol is available on a creative commons license and is free for organisations to apply.

Internationally Agreed Standard

Environmental-economic accounts provide the conceptual framework for integrated statistics on the environment and its relationship with the economy, including the impacts of the economy on the environment and the contribution of the environment to the economy. A coherent set of indicators and descriptive statistics can be derived from the accounts that inform a wide range of policies, including, but not limited to, green economy/green growth, natural resource management and sustainable development. The System of Environmental-Economic Accounting (SEEA) contains the internationally agreed standard concepts, definitions, classifications, accounting rules and tables for producing internationally comparable statistics on the environment and its relationship with the economy. The SEEA is a flexible system in the sense that its implementation can be adapted to countries' specific situations and priorities. Coordination of the implementation of the SEEA and on-going work on new methodological developments is managed and supervised by the UN Committee of Experts on Environmental-Economic Accounting (UNCEEA). The final, official version of the SEEA Central Framework was published in February 2014.

Natural Capital Accounting

Natural Capital accounting is a tool to measure the changes in the stock of natural capital at a variety of scales and to integrate the value of ecosystem services into accounting and reporting systems at Union and national level.

Methods of Accounting

There are several methods of accounting which aim to address the issue of sustainability. These are: large and eclectic dashboards; composite indices; indices focusing on overconsumption; adjusted economic indicators.

Large and eclectic dashboards

These dashboards bring together a number of indicators that are directly and indirectly related to the durability of socio-economic progress. One example of this is the Eurostat Sustainable Development Indicators, which is a list of over 100 indicators used to monitor the EU Sustainable Development Strategy. The criticism associated with these dashboards is that a large number of indicators risks muddling a clear message about sustainability that resonates with policy makers or citizens. In response, there has been a greater tendency to select headline indicators that "track central elements of green growth and representative of a broader set of green growth issues."

Composite Indices

Composite indices normalize and aggregate various data into a single number. For example, the Human Development Index, Osberg and Sharpe's Index of Economic Well-Being, the Changing Wealth of Nations, or the Environmental Sustainability Index, which ranks countries based on an assessment of 76 variables covering 5 domains. It is often instructive to examine the separate dimensions of these indices. However, they may present a skewed view of countries' contributions to environmental problems and make problematic, normative assumptions about the values of certain variables.

Adjusted GDPs

Adjusted gross domestic product, or green GDP, systematically corrects conventional GDP by taking into account aspects of a country's production of goods and services (e.g. environmental degradation and natural resource depletion) that would not otherwise be included in the indicator, but are relevant to sustainability.

Indices Focusing on Overconsumption

Indicators that fall in this category conceive of sustainability with respect to consumption levels and investment in natural resources. Examples include adjusted net savings (ANS) and ecological footprint accounts. ANS is calculated as the change in total wealth over a given time period, while ecological footprint assessments determine how much of the regenerative capacity of the biosphere is required to maintain the consumption habits of a defined population. The explicit emphasis on sustainability makes these indices useful; however computing them by country fails to capture the global nature of sustainability.

Monetary or Physical Indicators

All sustainability indicators can be grouped broadly into two types. Specifically, they will be calculated in monetary terms, using one or more valuation techniques, or in physical terms. It is more likely for monetary indicators to be expressed as flows, and physical indicators as stocks.

Global Initiatives

General commitment by the international community to support the development of natural capital accounting was motivated early on by the Brundtland Report in 1987 and the 1992 Rio Summit.

System of Environmental-economic Accounting

In September 1992, the Commission on the Environment of the Organization of American States (OAS) Permanent Council held a Seminar on Natural Resource and Environmental Accounts for Development Policy. Many of the country participants expressed interest in developing accounting capacities for natural resources. A proposal was made at that time to create a program to coordinate and strengthen the efforts of countries and institutions undertaking such initiatives.

The development of the first system of environmental-economic accounting (SEEA) in 1993 (SEEA-1993) was a major step towards establishing standards around integrating the environment into national accounts, and subsequently, environmentally-adjusting or "greening" macroeconomic indicators such as GDP. While the SEEA-2003 and subsequent revisions being undertaken for 2013 have expanded the range of analyses within the framework, the purpose of the SEEA has remained the same. It is an accounting framework that records the stocks and flows that are relevant to both the environment and the economy. Its Central Framework comprises three main accounts that can be integrated with the existing United Nations System of National Accounts (SNA), and each focuses on a different aspect of the interaction between the economy and the environment: physical flow accounts; functional accounts for environmental transactions; and asset accounts in physical and monetary terms.

The latest version of the SEEA (Q3, 2012) has two other parts, aside from the Central Framework: SEEA Experimental Ecosystem Accounts and SEEA Extensions. The Experimental Ecosystem Accounts, specifically, introduces an accounting framework for ecosystems, despite the fact that many of its relevant stocks and flows are centered on non-market assets. While some of the measurement concepts involved in the accounting process are still evolving, it is possible that the eventual valuation of ecosystems and their depletion could be included in the calculations of environmentally-adjusted macroeconomic indicators. This has implications for future policy, since the emphasis on certain projects or activities undertaken by governments will likely change, depending on how the above-mentioned measurements impact their respective accounts, and subsequent environmental adjustments to certain indicators.

The London Group on Environmental Accounting and the UN Committee of Experts on Environmental-Economic Accounting are two groups, created in 1994 and 2005 respectively, to assist in the development of the SEEA and its implementation. As well, the Working Group on Environmental Auditing, a subgroup of the International Organization of Supreme Audit Institutions, is working to improve auditing standards related to environmental issues.

Wealth Accounting and Ecosystem Partnership Services

The ability of developing countries to build their natural capital account capacities is being improved significantly through the Wealth Accounting and Ecosystem Partnership Services (WAVES), by encouraging the development of relevant measurement frameworks. WAVES is global partnership that was inaugurated in October 2010 by World Bank President Robert Zoellick at the Convention on Biological Diversity COP-10 meeting in Japan. It aims to promote sustainable development by encouraging the inclusion of natural capital measurements in national accounts. Several projects have been initiated in developing countries such as Botswana, Colombia and Madagascar with a view to improving their capacity to implement the SEEA, in collaboration with UNEP, the UNDP, the United Nations Statistical Commission, and the financial support of NGOs and the governments of Australia, Canada, France, Japan, Norway and the United States.

Ecological Footprint Accounts

Alternately, there have been many attempts to move away from integrated accounts, and towards novel sustainability indices or statistics. Ecological footprint accounts, developed by Monfreda (2004) and since 2007, the Global Footprint Network, or the proposal for nine planetary boundaries within which humans can safely operate, by Rockstrom, are projects which advocate for new approaches to global sustainability.

Inclusive Wealth Index

The UN International Human Dimensions Programme has created an inclusive sustainability indicator, the Inclusive Wealth Index (IWI), which measures the productive bases of an economy: produced, natural and human capital, and based on these three assessments, calculates the trajectory of a country's wealth. The calculation of natural capital in the IWI is based on the shadow value of an economy's natural capital assets.

A similar conceptual direction was taken by the Commission on the Measurement of Economic Performance and Social Progress, under the direction of economists Joseph Stiglitz, Amartya Sen and Jean-Paul Fitoussi, at the behest of former French President Nicholas Sarkozy, in 2008. The authors concluded that a pragmatic measure of sustainability would combine an indicator based on the extended wealth approach, and a small dashboard of physical indicators.

While they have different theoretical underpinnings, what these approaches have in common with each other is a fundamental recognition of the limitations of traditional indicators in measuring economic performance and social progress, and the importance of sustainability in the long-run.

Beyond GDP

There have been several initiatives organized at the regional level that seek to move away from traditional GDP as the major indicator of wealth and well-being. The first of these is Beyond GDP, an initiative started by the EU in 2007 to develop highly aggregated environmental and social indicators and extend the national accounts to environmental and social issues.

In May 2012, the Summit for Sustainability in Africa was held in Gaborone, Botswana. It also brought together leaders from ten African countries to discuss sustainable development planning, and in particular, to commit to a set of goals on fully accounting for natural capital, and integrating it into national planning, reporting and policies. This commitment took the form of the *Gaborone Declaration for Sustainability in Africa*

Economics of Ecosystems and Biodiversity

Four publications were presented at the CBD COP-10 by The Economics of Ecosystems and Biodiversity (TEEB) initiative: TEEB Ecological and Economic Foundations, TEEB in National and International Policy Making, TEEB in Local and Regional Policy, and TEEB in Business and Enterprise. These, along with an Interim Report released at the CBD COP-9 in Bonn, Germany, represent a comprehensive analysis of the economic value of biodiversity, and the consequences it holds for different levels of public and private policy analysis. TEEB also seeks to systematize the role of corporations, under the TEEB for Business Coalition (2012), by formulating standards and assessing externalities. Since natural capital accounting requires the identification of replenishment activities as well as environmental degradation, the inclusion of corporations into the valuation process is key.

Natural Capital Declaration

In June 2012, the Rio+20 conferences "marked a watershed in the world wide interest on Natural Capital Accounting". The Natural Capital Declaration (NCD), a commitment by CEOs in the financial sector to embed ESG considerations in management and investment activities, was revealed prior to the conference. As well, the World Bank started the WAVES 50:50 Initiative to analyze the progress and next steps required for improving efforts to account for natural capital and enhance countries' sustainable decision-making capabilities. At the time of the conference, 62 countries, 90 corporations, and 17 civil society members had signed on to the campaign.

Country Initiatives

Many countries are undertaking projects to develop environmental accounts, to integrate them with national accounts or to create environmentally-adjusted macroeconomic indicators, including Green GDP. Early adopters of integrated environmental-economic accounts include the Netherlands, France, and the Philippines.

Argentina

A team of researchers at the University of Buenos Aires and CONICET organized the AR-KLEMS+LAND project. Their research, based on the KLEMS database framework (Capital, Labour, Energy, Material and Service Inputs), measures and analyses the sources of economic growth, productivity and competitiveness in the Argentinian economy. The database includes the contributions of "Natural Resource as Land and Subsoil Assets" services to GDP growth.

Professor George Santopietro, at Radford University in Virginia, examined several alternative methods for estimating resource rent and relatedly, depletion costs of natural capital. The methods he engages with are: net price, El Serafy's depletion cost, sustainable price, transaction value and replacement cost. He derives data from the privatization of Yacimientos Petroliferos Fiscales (YPF), Argentina's state-owned oil enterprise, and using each method, generates estimated resource rent and depletion cost values for Argentina's reserves of petroleum. The results show that the net price and transaction value methods overvalue the resource rent of petroleum reserves. He concludes that rent should be derived using the value of a firm's stock.

Australia

Several Australian organizations produce national environmental accounts. The Australian Bureau of Statistics produces Water Account Australia and a pilot Land Account for the Great Barrier Reef, both of which are based on the SEEA framework. The Bureau of Meteorology produces a complementary National Water Account on the water available to be used. The Department of Climate Change and Energy Efficiency reports on greenhouse gas (GHG) emissions. The Wentworth Group of Concerned Scientists has created a trial, environmental accounting model that can be applied to regions of any size. These regional accounts use a common unit of measurement, which allows comparisons to be made between different natural capital assets. The Victorian Department of Sustainability and Environment is in the process of developing experimental ecosystem accounts.

The energy and water accounts found on the National Balance Sheet are produced annually. As well, subsoil assets, timber for producing logs, and land are valued monetarily, and included on the non-produced assets in the National Balance Sheet. The Australian Bureau of Statistics (ABS) is currently working with several other national departments on the National Plan for Environmental Information (NPEI), the State of the Environment Report, and planning for national environmental accounts.

In 2012, following the adoption of the SEEA as an international statistical standard, the Australian Bureau of Statistics (ABS) also published a report titled "Completing the Picture – Environmental Accounting in Practice," which develops a number of accounting tables rooted in SEEA formatting, and explores how these could influence policy decisions.

Britain

The UK produces Environmental Accounts, consistent with the SEEA framework, that are separated into three dimensions: natural resource accounts, physical flows and monetary accounts. The majority of the data is measured in physical units; monetary units are used where relevant, and if the necessary data is available.

The current environmental asset accounts produced by the Office for National Statistics (ONS) are: oil and gas reserves, forestry and land cover. In 2012, the ONS planned to run a study on producing a full ecosystem account, with high priority on physical and monetary forestry accounts. The Office also set out a series of goals to include land use and cover accounts in the 2013 Environmental Accounts.

The UK National Ecosystem Assessment (NEA) was a two-year initiative from 2009 to 2011 that assessed the benefits of the natural environment to society and the economy.

The Government has pledged to develop full UK Environmental Accounts by 2020. This work is being led by the Office for National Statistics (ONS) and the Department for Environment, Food and Rural Affairs (Defra). The UK Natural Capital Committee is providing advice to help inform this project. In December 2012, the ONS published a 'roadmap' that set out the work-plan for the project. The incorporation of natural capital into the national accounts, if done properly, would enable a high level picture to be obtained of trends in the state of the nation's natural assets through time. This in turn would help demonstrate the implications of actions impacting on the natural environment and may, therefore, encourage policy-makers to ensure natural assets are in future used sustainably through the appropriate use of policy levers.

The Natural Capital Committee is also developing a methodology for corporate natural capital accounting. It is undertaking a pilot project throughout 2014 with a range of organisations to test and refine this methodology and is encouraging organisations that own or manage land to develop corporate natural capital accounts.

Brazil

Despite reports that Brazil would produce a Green GDP in the lead-up to the Rio+20, there is no indication that this process was undertaken.

Canada

Statistics Canada does not redefine or supplement existing national accounts aggregates with environmental information. However, it has developed the Canadian System of Environmental and Resource Accounts (CSERA), which "provides some information for those who may wish to calculate such "green aggregates."" It also includes estimates of land in the "tangible non-produced assets" section of the Canadian balance sheet.

The CSERA has three dimensions: Natural Resource Stock Accounts; Material and Energy Flow Accounts; and Environmental Protection Expenditure Accounts. The Natural Resource Stock Accounts are measured in physical and monetary units. These, in turn, comprise natural resource wealth estimates that are included on the National Balance Sheet Accounts.

In 2005, the Pembina Institute published a report on the Canada Boreal Initiative, which sought to quantify the full economic value of goods and services provided by Canada's Boreal region. The Boreal Ecosystem Wealth Accounting System (BEWAS) was constructed for this purpose. It considered the physical conditions in the boreal region by using physical inventory and spatial data. The two-year study approximated the market value, for 2002, of the region's forestry, mining, oil and gas, and hydroelectric generation sectors at $48.9 billion. For the same time period, it also estimated the net market value of natural capital extraction at $37.8 billion and the non-market value of ecosystem services at $93.2 billion.

Professor Nancy Olewiler, at Simon Fraser University, conducted several case studies to value natural capital in settled areas of Canada. Her agricultural lands case studies found the total net value of conservation efforts to be approximately $195/ha/yr in the Ontario Grand River Watershed, $65/ha/yr in the Upper Assiniboine River Basin and $126/ha/yr in the Prince Edward Island Mill River Watershed.

China

In 1997, Beijing authorities carried out a project assessing the city's Green GDP. This set a series of precedents for environmentally adjusted indicators in the country. Over the next decade, several regional pilot projects were undertaken by local authorities. Between 2001 and 2004, Chinese authorities worked with Statistics Norway to carry out a Green GDP assessment of Chongqing.

Then, in 2004, the Green GDP Accounting Research Project was launched by the State Environmental Protection Administration of China (SEPA) and the National Bureau of Statistics (NBS). The findings, released in the China Green National Accounting Study Report 2004 in 2006, reported that environmental pollution cost the economy 511.8 billion yuan or 3.5% of GDP in 2004. A breakdown of the figure shows that water pollution, air pollution, and solid waste and accidents cost 286.28 billion yuan, 219.8 billion yuan, and 5.74 billion yuan, respectively. According to the report, only about ten of the items the project intended to distinguish were accounted for. The costs of resource depletion and ecological damage were not included in the calculations because of methodological difficulties, limited technological capabilities and a lack of relevant data.

At the same time as the report was released, SEPA Vice Minister Zhu Guangya issued an independent report stating that each year, environmental damage in the country cost approximately 10% of GDP. This estimate was consistent with the one that scientists, economists and the World Bank had expected, of 8-12% of GDP.

The government withdrew its official support for Green GDP in 2007, after early results showed the reduction of growth rates in some provinces to nearly zero. It also did not release an official Green GDP report for 2005, which had been scheduled for publication in March 2007.

Reports have periodically surfaced about re-calculating China's green GDP. In 2007, the All-China Environment Federation (ACEF) called on the National Development and Reform Commission (NDRC) to carry out a national accounting system. *China Daily* reported that the Ministry of Environmental Protection valued the cost of pollution to the economy at 1.4 trillion yuan in 2009.

Starting in 1998, the NBS developed rudimentary environmental accounts in forestry and energy. Since then, the NBS has expanded its environmental accounts to include pollution treatment, water, and minerals, along with the development of a comprehensive SEEA for China.

European Union

The interaction between Eurostat and national statistical offices of EU Member States was formalized in 2011, by the adoption of Regulation No 691 on European Environmental Economic Accounts. It requires Member States to report data and accounts on air emissions, taxes related to the environment, and material flows from 2012. Eurostat is also constructing environmental accounts expressed in physical and monetary terms, and asset accounts, as a step towards developing a regional SEEA.

The creation of inclusive wealth indicators is also a recognized priority of the EU. The proposed 7th Environment Action Programme (EAP) of the EC explicitly identifies this issue, by calling for further development and integration of economic and environmental indicators. The Shared Environmental Information System (SEIS) proposes to streamline the collection of data required for designing environmental indicators. The EU also committed to the Aichi Accord at the CBD COP-10, in which Parties agree to integrate biodiversity into their national accounts. This is on top of the experimental ecosystem accounting framework that was launched in 2009. The European Environment Agency (EEA) proposed that given the compatibility of the design with the SNA, it would be possible to use one particular indicator, Consumption of Ecosystem Capital (CEC) to adjust National Accounts aggregates to create, in particular, CEC Adjusted Net Domestic Product and CEC Adjusted Net National Income.

The EC Communication Roadmap to a Resource Efficient Europe sets 2020 as the year by which businesses, along with public authorities, will properly account for natural capital and ecosystem services.

France

The French Ministry of Sustainable Development produces satellite, environmental-economic accounts each year. However, it is planning on fully expanding the accounts to correspond with the SEEA (2012). The National Institute of Statistics and Economics also includes estimates of land, subsoil assets, and non-cultivated bio and water resources in the "tangible non-produced assets" section of the French balance sheet.

From 2008 to 2010, an exploratory study on ecosystem valuation was carried out. In accordance with the adoption of the Aichi Accord at the CBD COP-10, the French National Biodiversity Strategy

Target 7 was set to "include preservation of biodiversity in economic decisions." The Ministry of Sustainable Development is presently consulting with stakeholders and refining methodological options in order to begin assembling physical ecosystem assessments. Annual expert workshops on monetary valuation and economic instruments are also organized by the Ministry.

Germany

The German Environmental Economic Accounts (GEEA) follows the SEEA framework. The data from these accounts, along with the German Socio-economic Accounts, is used to calculate an indicator set, as outlined in the National Strategy for Sustainable Development (2002). While no adjusted macroeconomic aggregates are estimated in the GEEA, two of the GEEA indicators are embedded into the National Accounts: productivity of energy and raw materials, and transport intensity and share of the railways in providing transport.

India

An Expert Group, led by IHDP Scientific Committee chairman Professor Partha Dasgupta, is in the process of developing a system to "green" India's national accounts. It plans to adjust GDP to account for environmental costs and impacts by 2015. Some of the issues the Group will address include establishing coordination mechanisms within the country and with international partners, and standardizing data collection and valuation methodologies.

The Green Accounting for Indian States Project (GAISP) was the first initiative of the Green Indian States Trust (GIST), an NGO started in 2004. The Project used data from Indian national databases to measure sustainable development and creates green accounts at the state-level that were consistent with SEEA-2003 guidelines. Following this, GIST calculated a Green GDP total which adjusted the traditional indicator for "all major externalities."

The South Asian Network for Development and Environmental Economics (SANDEE) and the Indian Society for Ecological Economics operate in India, although SANDEE is based in Nepal. Both are networks that conduct research on and analyses of environmental accounting, which is instructive for teachers, researchers and members of the policy community.

Indonesia

Members of the World Resources Institute, led by David Repetto, integrated environmental effects into Indonesia's national accounts in 1990. He subtracted net natural resource depreciation for the petroleum, timber, and soils sectors from GDP to estimate the environmentally-adjusted NDP. They found that estimates of net income and growth of net income were overstated when calculated using the conventional accounts.

In 1997, the United Nations University-Institute of Advanced Studies (UNU-IAS) measured the environmental impacts of industrialization and trade in Indonesia (along with China and Japan). In their study, the UNU-IAS constructed an international environmental input-output model for the Asia Pacific region, and from this, was able to compile a preliminary SEEA and Green GDP approximation. In 1990, environmental costs were equal to 4.9% of NDP for Indonesia. In particular, changes in land use and oil exploitation were major costs.

Since 2002, Statistics Indonesia (BPS) has produced several Indicators of Sustainable Development reports, based on the 134 UN Commission on Sustainable Development Indicators developed by the UN Department of Economic and Social Affairs. BPS also regularly generates environmental statistics and statistics of marine and coastal resources.

Italy

The National Institute of Statistics (ISTAT) currently constructs three types of environmental accounting modules: material flows (MFA), the environmental account matrix integrated with national economic account (NAMEA), and the expenditure estimate for environmental protection (SERIEE-EPEA).

In 1999, Silvia Teizzi at the University of Sienna applied a method of monetary valuation to externalities arising from agricultural production in Italy. She separately estimated a shadow price and a quantity for the externalities and calculated their values between for each year between 1961 and 1991. She subtracted these figures from the value added of the agricultural sector "as a first step towards the correction of national accounting aggregates to take environmental degradation into account."

Under the local Agenda 21 process, the City and Local Environmental Accounting and Reporting (CLEAR) project and the CONTAROMA project were undertaken by Italian municipalities. The CLEAR project was started in 2001 to pair the financial statements of municipalities with an environmental report. The CLEAR method reclassified the aforementioned financial statements such that expenditures for environmental purposes could be identified, and compiled into separate monetary accounts. Similarly, the CONTAROMA project developed an environmental accounting system for the municipal budget of Rome.

Japan

In 1995, the Japanese Economic Planning Agency made initial estimates of the SEEA and Green GDP from 1985 to 1990. In 1998, they extended their estimates from 1970 to 1995. According to these calculations, environmental costs in Japan fell from 8% of NDP in 1970 to nearly 1% in 1995.

The E-10 is the official economic-environmental accounts produced by the Japanese Ministry of the Environment. They are based on data compiled by 10 government ministries, and describe the environmental burden of particular natural resources in physical units. There are three tables which make up the E-10: a Basic Transaction Table; resource and environmental burden tables; and supplementary tables.

The Statistics Bureau of Japan currently includes estimates of land in the "tangible non-produced assets" section of the Japanese balance sheet.

Republic of Korea

In 2001, the Korean Ministry of the Environment planned to introduce an environmentally-adjusted GDP indicator. It began developing a SEEA (KORSEEA) in 2002, and the development of Economy-wide Material Flow Accounts, and NAMEA took place in the following years. Several government institutions and departments provide economic and environmental data required for the KORSEEA. Statistics Korea also includes estimates of land, subsoil assets, and non-cultivated bio and water resources in the "tangible non-produced assets" section of the Koreah balance sheet.

The Korea Rural Economic Institute calculated the Green GDP of the Korean agricultural sector in the period 1980 to 1997 using pilot economic-environmental accounts. Environmental adjustment in this study is limited to the subtraction of "degradation costs of natural resources" from NDP. They report that the ratio of environmentally adjusted domestic product to NDP decreases from 100.6% to 99.5% in the agricultural sector during the specified period.

Mexico

In 1985, the UN, World Bank and Mexican government executed a joint pilot project using 1985 data to produce two environmentally-adjusted indicators: resource depletion subtracted from NDP and environmental degradation subtracted from NDP.

The Mexican System of Economic and Ecological Accounts (MSEEA) was started in 1988, and has been published annually since 1991. The National Institute of Statistics and Geography of Mexico (INEGI) have used the accounts to derive key information about environmental sustainability in the country. Yearly environmental costs are equal to 8.5% of GDP. Since 2003, the INEGI has calculated a yearly Green GDP aggregate. Between 2003 and 2009, the ratio of Green GDP to GDP increased from 90.4% to 92.1%. The INEGI reports that while the scope of action should be increased, this trend reflects the successful efforts of the government to reduce impacts that negatively affect the environment.

Russia

The Federal State Statistics Service (Rosstat) currently includes estimates of land and non-cultivated bio and water resources in the "tangible non-produced assets" section of the Russian balance sheet. Rosstat has outlined a series of priorities for developing a Russian SEEA. Its highest priority is asset accounts, and with respect to the environment, a particular record of environmental protection expenditures. After 2015, the basis for integration will appear with the creation of input-output tables. Physical flow accounts are a medium priority; further development will depend on the completion of the input-output tables and whether new surveys are introduced on time. While accounts for extended SNA aggregates, such as the depletion or degradation of natural capital are being given conceptual consideration, they are the lowest priority of Rosstat.

South Africa

The environmental-economic accounts are developed in the National Accounts Division of Statistics South Africa (SA). Information for the accounts is sourced from various government departments, surveys and geographic and spatial data. According to a 2011 report to the United Nations, environmental statistics are not a mainstream priority for the country, and Statistics SA does not have adequate resources to collect environmental data. Additionally, a "lengthy institutionalization process" has prevented the publication of any official reports.

Turkey

The Turkish Statistical Institute produces a range of environmental statistics

United States of America

In 1992, the Bureau of Economic Analysis within the Department of Commerce began intensively developing environmental accounting methodologies. However, Congress directed the Department to suspend all activities in this area in 1994, after the first U.S. Integrated Environmental and Economic Satellite Accounts were published. It also ordered an external of environmental accounting. This review, Nature's Numbers: Expanding the National Economic Accounts to Include the Environment was concluded in 1999.

William Nordhaus (Nature's Numbers), along with Nicholas Muller and Robert Mendelsohn, co-authored a study on integrating environmental externalities into a SNA. Their model estimates gross damages from air pollution in each industry in the United States. It estimates that damages from activities such as combustion of waste, sewage treatment and firing power plants by oil or coal are larger than their value added. The largest externalities, for which damages range from 0.8 to 5.6 times value added, arise from coal-fired electric generation.

Carrying Capacity

Carrying capacity is the number of organisms that an ecosystem can sustainably support.

An ecosystem's carrying capacity for a particular species may be influenced by factors as its ability to regenerate the food, water, atmosphere, and other necessities that the organisms need to survive.

Ecosystems cannot exceed their carrying capacity for a long period of time. In situations where the population density of a given species exceeds the ecosystem's carrying capacity, the species will deplete its own source of food, water, or other necessities, and will soon begin die off. The species population can grow until it reaches that line, at which point resources will not be sufficient to allow it to continue to grow over the long-term:

An example of a situation in which the carrying capacity of an environment is exceeded is the example of deer in North America.

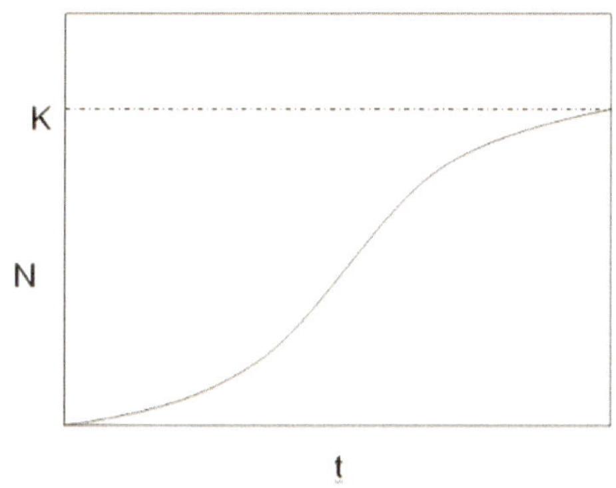

After the widespread elimination of wolves – the natural predator of North American deer – the deer reproduced until their need for food plant species exceeded the environment's ability to regenerate their food plants. In many areas, this resulted in large numbers of deer starving until the deer population was severely reduced.

Humans have managed to increase the Earth's carrying capacity for human life using technological inventions such as agriculture and genetically modified crops, which have a greater ability to produce food in some environments than natural crops.

Still, some scientists fear that humans may exceed the Earth's carrying capacity for humans, and encourage the use of contraception to decrease birth rates in order to prevent human populations from exhausting their sources of food and other vital resources.

Examples of Carrying Capacity

The Daisyworld Model

The hypothetical "Daisyworld" model is a model developed by scientists to study how organisms chance their environment, and how ecosystems self-regulate.

In the original "Daisyworld" mathematical simulation, there were only two types of life forms: black daisies, which increase the environment's temperature by absorbing heat from the Sun (this is a real property of black materials), and white daisies, which decrease the environment's temperature by reflecting the Sun's heat (this is also a real effect of white-colored materials).

Each species of daisies had to live in a proper balance with the other species. If the white daisies overpopulated, the world would become too cold. Daisies of both types would begin to die off, and the world would start to regain its pre-daisy equilibrium. The same held true for black daisies: if they overpopulated, the world would become warmer and warmer until the daisies began to die off again.

Real-life ecosystems are much more complicated than this, of course.

Each organism has many needs, and how well the environment can meet those needs might depend on what other organisms it shares the environment with. Such is the case for...

The North American Deer

Before Europeans colonized North America, one of its main forest herbivores were deer. Ordinarily existing in small groups, populations of deer were kept in check by wolves, the top predator of these forest ecosystems.

Deer, being fairly large as North American herbivores go, were capable of eating leaves off of trees and shrubs, as well as low-growing plants like flowers and grass. And they required a lot of leaves to keep them going, as members of different species of deer could weigh anywhere from 50 to 1,500 pounds.

But when European settlers severely depleted the population of wolves, who they found to be a danger to human children and livestock, an unexpected consequence resulted: deer began to multiply out of control, until they exceeded the carrying capacity of their environment.

As a result, deer began to starve. Plants species also began to suffer, some even being threatened with extinction as the starving deer ate all the green plants they could find.

When humans realized what was happening – and it began to affect their own food sources, after wild deer began to invade gardens and farms looking for crops to eat – they began to give nature a helping hand in reducing the deer population.

In modern times, some areas "cull" deer – a practice where deer are systematically hunted, not just for meat or sport, but to prevent deer starvation and damage to plants. Other areas have even begun to re-introduce wolves – humans' old enemy – and have seen healthier ecosystems, gardens, and crops as a result.

The story of the North American wolves and deer has acted as a cautionary tale for people considering making changes of any kind to their natural environment, which might have unintended consequences.

Humans

Humans have become one of the world's only global species my mastering technology. Time and time again, the human species has overcome a factor, such as availability of food or presence of natural predators that limited our population.

The first major human population explosion happened after the invention of agriculture, in which humans learned that we could grow large numbers of our most nutritious food plants by saving seeds to plant in the ground, and making sure those seeds got enough water, were protected from competition from weeds and from being eaten by other animals, etc..

When agriculture was invented, the human population skyrocketed – scientists think that without agriculture, between 1 million and 15 million humans were able to live on Earth at a time. Today, there are about 1 million humans in the city of Chicago alone.

By the Middle Ages, when well-organized agriculture had emerged on every continent, there were about 450 million – or about half a billion – humans on earth.

A new revolution in Earth's capacity to carry humans began in the 18th and 19th centuries, when humans began to apply advanced and automated technology to agriculture. The use of inventions such as the mechanical corn picker and crop rotation – a way of growing different crops in sequence that enriches the soil and leads to higher yields – allowed humans to produce even more food. As a result, the world population tripled from about half a billion to 1.5 billion people.

In the twentieth century, a third revolution occurred when humans began to learn how to rewrite the genomes of the plants, using viruses to insert new genes into seeds directly instead of relying on selective breeding and random mutation to increase crop yields. The result was another drastic increase in the Earth's ability to produce food for humans.

During the 20th century, Earth's human population more than quadrupled, from 1.5 billion to 6.1 billion. We've come a long way from the pre-agricultural days.

But some scientists worry that we may be well on our way to exceeding the Earth's carrying capacity – or that we may have already done so.

Scientists point out that the practices currently used to grow genetically modified crops may not be sustainable, as they often require the use of chemicals that can harm other parts of our ecosystems.

Scientists point to the rapid decline of bee populations – which are necessary to pollinate some of our crops, and which many scientists believe are being killed by pesticides we use to protect those same crops – as evidence that our current food production practices may not be sustainable for much longer.

The proliferation of poisonous algae, which can poison our water supplies and which feeds on the same fertilizer we use to feed our crops, is another worrisome sign that we may be exceeding our carrying capacity, and may begin to cause problems for ourselves if our population continues to grow.

Some believe that humans will continue to invent newer, better ways to grow food and to make water safe to drink. Many scientists agree that this is true: however, they caution that we should not simply continue to grow our population and assume that advances in technology will be able to take care of everyone.

Instead, scientists urge us to consciously plan our birth rates and population growth to avoid humanitarian crises.

Degrowth

The concept of degrowth has been described as an equitable downscaling of production and consumption that increases human well-being and enhances ecological conditions at the local and global level, in the short and long term. Critiques of growth began to be vocalised in the environmental and counterculture movements of the 1960s and 1970s. With French roots, 'Decroissance' was coined in 1979 by Nicholas Georgescu-Roegen in a book entitled Demain la decroissance. Other influential works critical of economic growth were published in the 1970s: Ivan Illich, Barry Commoner and André Gorz, are well-known authors of that decade. Essentially five different and overlapping sources are identified as having driven the conception and development of degrowth: ecological economics/bioeconomics; ecologists/environmentalists; cultural diversity/post-development; democracy/critical politics and spirituality/voluntary simplicity.

Degrowth has now become a political, economic and social movement based on environmentalist, anti-consumerist and anti-capitalist ideas. It can be described as a galaxy of people willing to experiment with or advocate alternative ways of co-existing with the goal of maximizing happiness and well-being through non-consumptive means: reducing work time, consuming less, while devoting more time to art, music, family, culture and community. These experiments occur at three levels: individual, collective/communal and political. At the individual level, degrowth is achieved by voluntary simplicity. Conviviality and slowness are endorsed through collective projects. Proposals for global solutions involve the relocalisation of economic activities in order to end humanity's dependence on fossil fuels and reduce its ecological imprint.

Despite improvements in the ecological efficiency of the production and consumption of goods and services, global economic growth has led to increased extraction of natural resources and waste/emissions. Moreover, unequal exchange in trade and financial markets has increased inequality between countries.

Resource Depletion

As economies grow, the need for resources grows accordingly. There is a fixed supply of non-renewable resources, such as petroleum (oil), and these resources will inevitably be depleted. Renewable resources can also be depleted if extracted at unsustainable rates over extended periods. For example, this has occurred with 'caviar' production in the Caspian Sea. There is much concern as to how growing demand for these resources will be met as supplies decrease. Many organizations and governments look to energy technologies such as biofuels, solar cells, and wind turbines to meet the demand gap after peak oil. Others have argued that none of the alternatives could effectively replace versatility and portability of oil. Authors of the book *Techno-Fix* criticize technological optimists for overlooking the limitations of technology in solving agricultural and social challenges arising from growth.

Proponents of degrowth argue that decreasing demand is the only way of permanently closing the demand gap. For renewable resources, demand, and therefore production, must also be brought down to levels that prevent depletion and are environmentally healthy. Moving toward a society that is not dependent on oil is seen as essential to avoiding societal collapse when non-renewable resources are depleted. "But degrowth is not just a quantitative question of doing less of the same, it is also and, more fundamentally, about a paradigmatic re-ordering of values, in particular the reaffirmation of social and ecological values and a repoliticisation of the economy".

Ecological Footprint

The ecological footprint is a measure of human demand on the Earth's ecosystems. It compares human demand with planet Earth's ecological capacity to regenerate. It represents the amount of biologically productive land and sea area needed to regenerate the resources a human population consumes and to absorb and render harmless the corresponding waste. According to a 2005 Global Footprint Network report, inhabitants of high-income countries live off of 6.4 global hectares (gHa), while those from low-income countries live off of a single gHa. For example, while each inhabitant of Bangladesh lives off of what they produce from 0.56 gHa, a North American requires 12.5 gHa. Each inhabitant of North America uses 22.3 times as much land as a Bangladeshi. According to the same report, the average number of global hectares per person was 2.1, while current consumption levels have reached 2.7 hectares per person. In order for the world's population to attain the living standards typical of European countries, the resources of between three and eight planet Earths would be required. In order for world economic equality to be achieved with the current available resources, rich countries would have to reduce their standard of living through degrowth. The eventual reduction of all available resources would lead to a forced reduction in consumption. Controlled reduction of consumption would reduce the trauma of this change assuming no technological changes increase the planet's carrying capacity.

Degrowth and Sustainable Development

Degrowth thought is in opposition to all forms of productivism (the belief that economic productivity and growth is the purpose of human organization). It is, thus, opposed to the current form of sustainable development. While the concern for sustainability does not contradict degrowth, sustainable development is rooted in mainstream development ideas that aim to increase capitalist growth and consumption. Degrowth therefore sees sustainable development as an oxymoron, as any development based on growth in a finite and environmentally stressed world is seen as inherently unsustainable. Critics of degrowth argue that a slowing of economic growth would result in increased unemployment, increase poverty and decrease income per capita. Many who understand the devastating environmental consequences of growth still advocate for economic growth in the South, even if not in the North. But, a slowing of economic growth would fail to deliver the benefits of degrowth—self-sufficiency, material responsibility—and would indeed lead to decreased employment. Rather, degrowth proponents advocate for a complete abandonment of the current (growth) economic system, suggesting that relocalizing and abandoning the global economy in the Global South would allow people of the South to become more self-sufficient and would end the overconsumption and exploitation of Southern resources by the North.

The Rebound Effect

Technologies designed to reduce resource use and improve efficiency are often touted as sustainable or green solutions. Degrowth literature, however, warns about these technological advances due to the “rebound effect”. This concept is based on observations that when a less resource-exhaustive technology is introduced, behaviour surrounding the use of that technology may change, and consumption of that technology could increase or even offset any potential resource savings. In light of the rebound effect, proponents of degrowth hold that the only effective ‘sustainable’ solutions must involve a complete rejection of the growth paradigm and a move toward a degrowth paradigm. There are also fundamental limits to technological solutions in the pursuit of degrowth, as all engagements with technology increase the cumulative matter-energy throughput. However, the convergence of digital commons of knowledge and design with distributed manufacturing technologies may arguably hold potential for building degrowth future scenarios.

Origins of the Movement

The contemporary degrowth movement can trace its roots back to the anti-industrialist trends of the 19th century, developed in Great Britain by John Ruskin, William Morris and the Arts and Crafts movement, in the United States by Henry David Thoreau, and in Russia by Leo Tolstoy.

The concept of “degrowth” proper appeared during the 1970s, proposed by André Gorz and intellectuals such as Nicholas Georgescu-Roegen, Jean Baudrillard, Edward Goldsmith and Ivan Illich, whose ideas reflect those of earlier thinkers, such as the economist E. J. Mishan, the industrial historian Tom Rolt, and the radical socialist Tony Turner. The writings of Mahatma Gandhi and J. C. Kumarappa also contain similar philosophies, particularly regarding his support of voluntary simplicity.

More generally, degrowth movements draw on the values of humanism, enlightenment, anthropology and human rights.

Lasting Influence of Georgescu-Roegen

The degrowth movement recognises Romanian American mathematician, statistician and economist Nicholas Georgescu-Roegen as the main intellectual figure inspiring the movement. In his magnum opus on *The Entropy Law and the Economic Process*, Georgescu-Roegen argues that economic scarcity is rooted in physical reality; that all natural resources are irreversibly degraded when put to use in economic activity; that the carrying capacity of Earth—that is, Earth's capacity to sustain human populations and consumption levels—is bound to decrease sometime in the future as Earth's finite stock of mineral resources is presently being extracted and put to use; and consequently, that the world economy as a whole is heading towards an inevitable future collapse.

Georgescu-Roegen's intellectual inspiration to degrowth goes back to the 1970s. When Georgescu-Roegen delivered a lecture at the University of Geneva in 1974, he made a lasting impression on the young and newly graduated French historian and philosopher Jacques Grinevald (fr), who had earlier been introduced to Georgescu-Roegen's magnum opus by an academic advisor. Georgescu-Roegen and Grinevald soon made friends, and Grinevald started devoting his research to a closer study of Georgescu-Roegen's work. As a result, in 1979 Grinevald published a French translation of a selection of Georgescu-Roegen's. Georgescu-Roegen, who spoke French fluently, personally approved the use of the term *décroissance* in the title of the French translation. The book gained influence in French intellectual and academic circles from the outset. Later, the book was expanded and republished in 1995, and once again in 2006; however, the word *Demain* ('Tomorrow') was removed from the title of the book in these second and third editions.

By the time Grinevald suggested the term *décroissance* to form part of the title of the French translation of Georgescu-Roegen's work, this term had already disseminated through French intellectual circles since the early 1970s to signify a deliberate political action to downscale the economy on a permanent and voluntary basis. Simultaneously, but independently hereof, Georgescu-Roegen had criticised the ideas of *The Limits to Growth* and Herman Daly's steady-state economy in his pointed and polemical.

Authors who were set exclusively on proving the impossibility of growth were easily deluded by a simple, now widespread, but false syllogism: Since exponential growth in a finite world leads to disasters of all kinds, ecological salvation lies in the stationary state. The crucial error consists in not seeing that not only growth, but also a zero-growth state, nay, even a declining state which does not converge toward annihilation, cannot exist forever in a finite environment.

The important, yet unnoticed point is that the necessary conclusion of the arguments in favor of that vision of a stationary state is that the most desirable state is not a stationary, but a declining one. Undoubtedly, the current growth must cease, may be reversed.

When reading this particular passage of the text, Grinevald realised that no professional economist of any orientation had ever reasoned like this before. Grinevald also realised the striking conceptual resemblance between Georgescu-Roegen's viewpoint and the French debates progressing by the time; this resemblance then transformed into the title of the French edition. Taken together, the translation of Georgescu-Roegen's work into French *both* fed on *and* gave further impetus to the concept of *décroissance* in the country—and everywhere else in the francophone world—thereby creating something of an intellectual feedback loop.

By the 2000s, when *décroissance* was to be translated from French and back into English as the catchy banner for the new social movement, the original term 'decline' was now deemed inappropriate and misdirected for the purpose: 'Decline' usually refers to an unexpected, unwelcome and temporary economic recession, something bad to be avoided or quickly overcome. Instead, the neologism 'degrowth' was coined to signify a deliberate political action to downscale the economy on a permanent and voluntary basis—as in the prevailing French usage of the term—something good to be welcomed and maintained, or so followers believe.

When the first international degrowth conference of its kind was held in Paris in 2008, the participants bestowed a generous amount of credit and appreciation on Georgescu-Roegen and his work. Further, in his manifesto on *Petit traité de la décroissance sereine* ('Farewell to Growth'), leading French champion of the degrowth movement Serge Latouche has credited Georgescu-Roegen as 'a main theoretical source of degrowth.' Likewise, Italian degrowth theorist Mauro Bonaiuti has considered Georgescu-Roegen's work to be 'one of the analytical cornerstones of the degrowth perspective.'

Serge Latouche

Serge Latouche, a professor of economics at the University of Paris-Sud, has noted that:

If you try to measure the reduction in the rate of growth by taking into account damages caused to the environment and its consequences on our natural and cultural patrimony, you will generally obtain a result of zero or even negative growth. In 1991, the United States spent 115 billion dollars, or 2.1% of the GDP on the protection of the environment. The Clean Air Act increased this cost by 45 or 55 million dollars per year. The World Resources Institute tried to measure the rate of the growth taking into account the punishment exerted on the natural capital of the world, with an eye towards sustainable development. For Indonesia, it found that the rate of growth between 1971 and 1984 would be reduced from 7.1 to 4% annually, and that was by taking only three variables into consideration: deforestation, the reduction in the reserves of oil and natural gas, and soil erosion.

Ecological and Social Issues

In January 1972 Edward Goldsmith and Robert Prescott-Allen published *A Blueprint for Survival*, which called for a radical programme of decentralisation and deindustrialization to prevent what the authors referred to as *"the breakdown of society and the irreversible disruption of the life-support systems on this planet"*.

Eco-restructuring

Eco-restructuring is the implication for an ecologically sustainable economy. The principle of ecological modernization establishes the core literature of the functions that eco-restructuring has within a global regime. Eco-restructuring has an emphasis on the technological progressions within an ecological system. Government officials implement environmental policies to establish the industrial- ecological progressions that enable the motion of economic modernization. When establishing economic growth, policy makers focus on the progression towards a sustainable environment by establishing a framework of ecological engineering. Government funding is necessary when investing in efficient technologies to stimulate technological development.

Environmental dimensions such as the preservation of landscape is achieved through sustainability. When analyzing the issues of sustainability, the importance of socio-technical transitions is highlighted to meet the needs of ecological policies. When an organization of resources is sustainable, it means that it will remain at a certain level of progress due to self-preservation without affecting the fundamental purpose of its existence. Changes in the industrial sector allows for sustainability to progress in an effective manner. When eco-restructuring is applied to the concept of sustainability, environmentally friendly technologies are essentially the main framework for clean and efficient productions.

Social support and acknowledgement must be present when highlighting the functions of modernizing the ecological structures. To have a smooth transition between socio-technical categories, social agreement is important when introducing eco-restructuring methods. When forming new ecological policies, policy makers focus on the movement that sustainable environments have in regards to the social expectations.

Economic production is an example of the existing relationship between eco-restructuring and sustainability. Cleaner production with minimal impact on the environment is the main idea behind the correlation of the two subjects. Since industrial ecology contributes to the direction of sustainability, eco-restructuring is perceived as a benefit for the environment and natural adaptation. Critics of the eco-restructuring and sustainability policies argue that a clear balance between a self-preserving ecological system and mass production is difficult to maintain. However, ecological engineering has been a major component used when applying eco-restructuring methods in the production of resources.

Technologies

Efficient technologies are defined as services that reduce the non-renewable resources produced and increase the production rate. Renewable energy is a form of energy obtained from clean and natural resources. Solar and wind power technologies are examples of clean energy forms that would eliminate pollution produced from the coal and diesel energy sectors.

Economic Movement

The economic movement outlined by Robert Ayers refers to the use of green economies. This type of economy structure looks at ways to reduce the environmental impacts. Nations focus on steering their policy agenda to allow the implementation of ecological strategies. This will accelerate the transformation of the economy sectors towards a clean and effective ecological state.

Social Lifestyles

Human economic activities can influence the type of issues that ecological systems face. These social lifestyles are subject to the changes in the production of goods and services. An example of this idea would be the collection of raw materials and disposing them in a proper efficient manner that will have little impact on the environment. If the whole cycle is restructured to meet the principles of ecological regulation, then experts would have to analyze the effects of human activities and determine a possible alternative to the removal of biomedical waste and other raw materials.

Many of the eco-restructuring principles were outlined in the *Eco-restructuring: Implications for Sustainable Development* book. However, policy makers can also have an opinion on the ideas and promises that this method of sustainability offers. Certain factors can influence the regulations that the policies cover such as the need for sustainable development. Legislative measures are also developed based on the propositions of ecological and economical advantages. The principle of sustainable development applies to the restructuring of technologies, economic movement and social lifestyles. These principles have helped to determine an evolution of the ecological system.

Environmental Factors

When an ecological system is altered, environmental standards must be put in place to preserve the stability of the environment. The application of eco-restructuring seeks to improve the environmental stability in relation to manufacturing and waste management. However, ecological effects can occur by following improper procedures. Despite the precautions that must be considered, there are environmental benefits of ecological modernization such as renewable energy. The use of renewable energy would eliminate the threat of ecological dangers such as pollution. Modern technologies also have a role in the impact of the environment. When new technologies are developed, sustainability is achieved through improving the environmental dimensions.

Eco-restructuring and Sustainability

Socio-technical transitions correlate to the need for social change in the study of sustainable resources. Sustainability has a significant role in the fields of management such as education, health care, economics, ecology and technology. The main framework for clean and efficient productions is supported by the idea of a self-preserving system. Despite the research supporting the reformation of the ecological system, natural resources can suffer depletion and will need the support of sustainability to preserve its existence. New policies are required to move towards equality and sustainability since the social expectations reflect the relationship between the two concepts. The public perception for an alternative solution to sustainable development is reflected through the ecological policies enacted by government officials. Eco-restructuring offers a positive aspect to the principles of ecological modernization. It is an alternative to the motion of technological progressions within the global economic regime.

Ecological Footprint

Ecological footprint (EF) is the measure of the demands made by a person or group of people on global natural resources. It has become one of the most widely used measures of humanity's effect upon the environment and has been used to highlight both the apparent unsustainability of current practices and the inequalities in resource consumption between and within countries.

The EF estimates the biologically productive land and sea area needed to provide the renewable resources that a population consumes and to absorb the wastes it generates—using prevailing technology and resource-management practices—rather than trying to determine how many people a given land area or the entire planet can support. It measures the requirements for productive areas (croplands, grazing lands for animal products, forested areas to produce wood products, marine areas for fisheries, built-up land for housing and infrastructure, and forested land needed to absorb carbon dioxide emissions from energy consumption). One can estimate the EF, measured in "global hectares" (gha), at various scales—for individuals, regions, countries, and humanity as a whole. (One hectare equals 2.47 acres.) The resulting figures can also be compared with how much productive area—or bio capacity—is available.

Canadian ecologist William Rees created the EF concept, which Swiss urban planner Mathis Wackernagel further developed in his dissertation under Rees's supervision. Together, Wackernagel and Rees wrote Our Ecological Footprint (1996), which describes the concept.

EF calculations have questioned the sustainability and equity of current consumption and production practices. The Global Footprint Network (GFN)—a nonprofit organization that partnered with hundreds of cities, businesses, and other entities to advance the EF as a metric of sustainability—calculates the per capita global footprint. In 2007 the per capita global footprint was 2.7 gha. Since global bio capacity in 2007 was 1.8 gha per person, the EF of humanity overshot Earth's bio capacity by 0.9 gha. In other words, 1.5 "Earths" would be needed to sustain current resource demands or, alternatively, it takes Earth one year and six months to regenerate what is used in one year. The implication of such ecological overshoot, which began in the mid-1970s, is that life-supporting biological resources, such as fisheries, forest resources, rangeland, and agricultural land, are being depleted.

Using the EF

EF analysis can show whether a country is living within the biocapacity of its own territory or whether it is an "ecological debtor," drawing on the ecological "capital" of other parts of the world. Per capita EFs show a wide divergence in the demands on nature from people in different societies, ranging from the United Arab Emirates at the high end (12.6 gha/person) to Haiti at the low end (0.5), with the United States (8.0), Germany (4.7), China (1.7), and others in between (2010 data). These figures are the basis of claims such that if all of humanity consumed like the average American, five Earths would be needed. EFs also vary greatly within countries according to level of affluence.

Researchers have combined footprint analysis with measures of human development to assess whether countries are on track toward sustainable development—defined as a per capita EF lower than the available per capita bio capacity with a high rating (above 0.8) on the United Nations Human Development Index (HDI). (The HDI is a metric that combines a country's average life expectancy, educational attainment, and income into a measure of economic and social progress).

Environmental educators and activists have used the EF to raise awareness of unsustainable consumption patterns, often with the goal of encouraging a change in lifestyles and, less frequently, to promote awareness of wider structural forces driving such patterns. Many online footprint calculators have appeared on nongovernmental organization Web sites with such goals in mind. Those calculators allow people to calculate their personal EF and to make comparisons with estimates of available bio capacity or to average footprints of other people locally and globally. Meanwhile, social scientists have used the EF as a comprehensive indicator of the ecological impacts of humans on the planet in order to test empirically different social theories of the forces driving those impacts.

Although EF analysis can lead to a radical critique of the current economic and social order, it has found increasing mainstream acceptance among businesses and governments. Multinational corporations such as Wal-Mart and BP introduced programs to shrink their EFs, and the World Business Council for Sustainable Development began exploring pathways toward achieving a "one-planet economy" by 2050. A number of national governments—such as those of Wales, Switzerland, the United Kingdom, Japan, the United Arab Emirates, and the European Union as a whole—as well as metropolitan areas have also considered or embraced the concept.

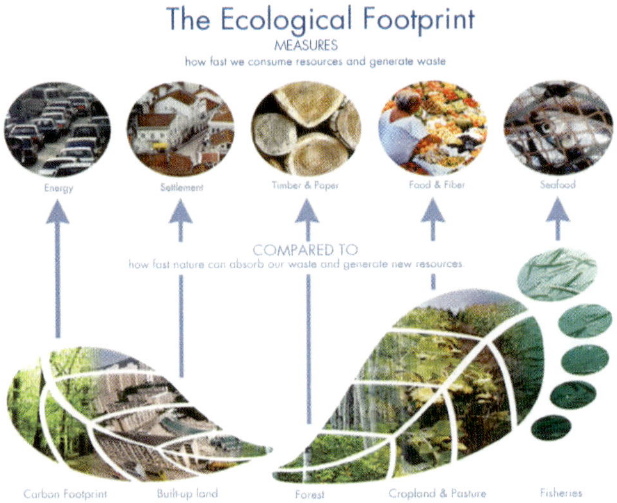

References

- Soderbaum, P. 2008. Understanding Sustainability Economics. Earthscan, London. ISBN 978-1-84407-627-7. pp.109-110, 113-117
- What-is-ecological-economics-one-pag-pdf-080123: sustainablepath.org, Retrieved 15 April 2018
- Costanza R; et al. (1998). "The value of the world's ecosystem services and natural capital1". Ecological Economics. 25 (1): 3–15. doi:10.1016/S0921-8009(98)00020-2.
- Encyclopedias-almanacs-transcripts-and-maps/green-politics, environment: encyclopedia.com, Retrieved 25 June 2018
- "Spash, C.L. (2011) Social ecological economics: Understanding the past to see the future. American Journal of Economics and Sociology 70, 340-375" (PDF). Archived from the original (PDF) on 2014-01-07. Retrieved 2014-01-07
- What-is-natural-capital: circulatenews.org, Retrieved 12 April 2018
- Stabile, Donald R. "Veblen and the Political Economy of the Engineer: the radical thinker and engineering leaders came to technocratic ideas at the same time," American Journal of Economics and Sociology (45:1) 1986, 43-44
- Socially-sustainable-economic-degrowth: ejolt.org, Retrieved 22 May 2018
- Knapp G, Roheim CA and Anderson JL (2007) The Great Salmon Run: Competition Between Wild And Farmed Salmon World Wildlife Fund. ISBN 0-89164-175-0

6

Ecological Restoration

Ecological restoration can be defined as the scientific study that supports the practice of renovation and restoration of degraded, destroyed or damaged ecosystems or habitats through active human intervention. This chapter discusses in detail the domains of ecological restoration such as resto-ration ecology, forest restoration, land restoration, etc.

Ecological restoration is the process of assisting the recovery of an ecosystem that has been degraded, damaged, or destroyed (Society for Ecological Restoration definition). Ecological restoration is a means of sustaining the diversity of life on Earth and re-establishing an ecologically healthy relationship between nature and culture. Ecological restoration can and should be a fundamental component of conservation and sustainable development programs throughout the world by virtue of its inherent capacity to provide people with the opportunity to not only repair ecological damage, but also improve the human condition.

Restoration Ecology

Restoration ecology refers to activities undertaken to increase populations of an endangered species or to manage or reconstruct a threatened ecosystem. Ecological restoration is an extremely difficult and expensive endeavour, and only attempted when the population of an endangered species is considered too small to be self-maintaining or the area of a threatened ecosystem is not large enough to allow its persistence over the longer term.

Restoration ecology can have various goals. A common focus is on endangered species and their habitat. In such a case, a species might be preserved in its remaining natural habitat, conserved by strictly controlling its exploitation, enhanced by a captive breeding and release program, and/or have its habitat managed to ensure its continued suitability. For example, the Galapagos Islands are maintained undisturbed by human development and multiple endangered species are thus, protected. In a more developed or suburban setting, this method is impractical and too expensive to implement. In the case of the California condor (Gymnogyps californianus), only few remaining natural habitats are carefully maintained, but the birds themselves are also monitored by visual observation and radio tracking. Several attempts have been made to increase the species population through breeding programs.

If a complement of species is being managed in some region, for example in a national park, the goal might focus on ensuring that all of the known native species are present and capable of sustaining their populations. If some species have been extirpated, there may be an effort to introduce new breeding populations. Habitat management might also be a component of this sort of multi-species goal.

If an endangered natural community is the focus, a project in restoration ecology might attempt to repair degraded remnants that still remain, or try to reconstruct a facsimile of the natural community. Restoration of the natural community may be accomplished by introducing native species missing from the ecosystem, and by managing the local environment to ensure the survival of all components of the community in an appropriate balance. The goal of community-level projects is to restore self-maintaining ecological communities that are as similar as possible to the original. This aspiration is rarely attainable to perfection, although it can be approximated to a significant degree.

Theoretical Foundations of Restoration Ecology

Restoration ecology draws on a wide range of ecological concepts.

Disturbance

Disturbance is a change in environmental conditions that disrupts the functioning of an ecosystem. Disturbance can occur at a variety of spatial and temporal scales, and is a natural component of many communities. For example, many forest and grassland restorations implement fire as a natural disturbance regime. However the severity and scope of anthropogenic impact has grown in the last few centuries. Differentiating between human-caused and naturally occurring disturbances is important if we are to understand how to restore natural processes and minimize anthropogenic impacts on the ecosystems.

Succession

Ecological succession is the process by which a community changes over time, especially following a disturbance. In many instances, an ecosystem will change from a simple level of organization with a few dominant pioneer species to an increasingly complex community with many interdependent species. Restoration often consists of initiating, assisting, or accelerating ecological successional processes, depending on the severity of the disturbance. Following mild to moderate natural and anthropogenic disturbances, restoration in these systems involves hastening natural successional trajectories. However, in a system that has experienced a more severe disturbance (i.e. physical or chemical alteration of the environment), restoration may require intensive efforts to recreate environmental conditions that favor natural successional processes.

Fragmentation

Habitat fragmentation describes spatial discontinuities in a biological system, where ecosystems are broken up into smaller parts through land use changes (e.g. agriculture) and natural disturbance. This both reduces the size of the populations and increases the degree of isolation. These smaller and isolated populations are more vulnerable to extinction. Fragmenting ecosystems decreases quality of the habitat. The edge of a fragment has a different range of environmental conditions and therefore supports different species than the interior. Restorative projects can increase the effective size of a population by adding suitable habitat and decrease isolation by creating habitat corridors that link isolated fragments. Reversing the effects of fragmentation is an important component of restoration ecology.

Ecosystem Function

Ecosystem function describes the most basic and essential foundational processes of any natural systems, including nutrient cycles and energy fluxes. An understanding of the complexity of these ecosystem functions is necessary to address any ecological processes that may be degraded. Ecosystem functions are emergent properties of the system as a whole, thus monitoring and management are crucial for the long-term stability of ecosystems. A fully functional ecosystem that is completely self-perpetuating is the ultimate goal of restorative efforts.

Community Assembly

Community assembly "is a framework that can unify virtually all of (community) ecology under a single conceptual umbrella". Community assembly theory attempts to explain the existence of environmentally similar sites with differing assemblages of species. It assumes that species have similar niche requirements, so that community formation is a product of random fluctuations from a common species pool. Essentially, if all species are fairly ecologically equivalent, then random variation in colonization, and migration and extinction rates between species, drive differences in species composition between sites with comparable environmental conditions.

Population Genetics

Genetic diversity has shown to be as important as species diversity for restoring ecosystem processes. Hence ecological restorations are increasingly factoring genetic processes into management practices. Population genetic processes that are important to consider in restored populations include founder effects, inbreeding depression, outbreeding depression, genetic drift, and gene flow. Such processes can predict whether or not a species successfully establishes at a restoration site.

Applications of Restoration Ecology

Soil heterogeneity Effects on Community Heterogeneity

Spatial heterogeneity of resources can influence plant community composition, diversity, and assembly trajectory. Baer manipulated soil resource heterogeneity in a tallgrass prairie restoration project. They found increasing resource heterogeneity, which on its own was insufficient to insure species diversity in situations where one species may dominate across the range of resource levels. Their findings were consistent with the theory regarding the role of ecological filters on community assembly. The establishment of a single species, best adapted to the physical and biological conditions can play an inordinately important role in determining the community structure.

Invasion and Restoration

Restoration is used as a tool for reducing the spread of invasive plant species in a number of ways. The first method views restoration primarily as a means to reduce the presence of invasive species and limit their spread. As this approach emphasizes control of invaders, the restoration techniques can differ from typical restoration projects. The goal of such projects is not necessarily to restore an entire ecosystem or habitat. These projects frequently use lower diversity mixes of aggressive native species seeded at high density. They are not always actively managed following seeding. The target areas for this type of restoration are those which are heavily dominated by invasive species. The goals are to first remove the species and then in so doing, reduce the number of invasive seeds being spread to surrounding areas. This approach has been shown to be effective in reducing weeds, although it is not always a sustainable solution long term without additional weed control, such as mowing, or re-seeding.

Restoration projects are also used as a way to better understand what makes an ecological community resistant to invasion. As restoration projects have a broad range of implementation strategies and methods used to control invasive species, they can be used by ecologists to test theories about invasion. Restoration projects have been used to understand how the diversity of the species introduced in the restoration affects invasion. We know that generally higher diversity prairies have lower levels of invasion. Incorporation of functional ecology has shown that more functionally diverse restorations have lower levels of invasion. Furthermore, studies have shown that using native species functionally similar to invasive species are better able to compete with invasive species. Restoration ecologists have also used the variety of strategies employed at different restoration sites to better understand the most successful management techniques to control invasion.

Successional Trajectories

Progress along a desired successional pathway may be difficult if multiple stable states exist. Looking over 40 years of wetland restoration data, Klötzli and Gootjans argue that unexpected and undesired vegetation assemblies "may indicate that environmental conditions are not suitable for target communities". Succession may move in unpredicted directions, but constricting environmental conditions within a narrow range may rein in the possible successional trajectories and increase the likelihood of a desired outcome.

Sourcing Material for Restoration

For most restoration projects it is generally recommend to source material from local populations, to increase chance of restoration success and minimize the effects of maladaptation. However the definition of local can vary based on species. Habitat and region. US Forest Service recently developed provisional seed zones based on a combination of minimum winter temperature zones, aridity, and the Level III Eco regions. Rather than putting strict distance recommendations, other guidelines recommend sourcing seeds to match similar environmental conditions. For example, sourcing for *Castilleja levisecta* found that farther source populations that matched similar environmental variables were better suited for the restoration project than closer source populations.

Principles of Ecological Restoration

Rationale

There are many reasons to restore ecosystems. Some include:

- Restoring natural capital such as drinkable water or wildlife populations

- Mitigating climate change (e.g. through carbon sequestration)
- Helping threatened or endangered species
- Aesthetic reasons
- Moral reasons: Human intervention has unnaturally destroyed many habitats, and there exists an innate obligation to restore these destroyed habitats.

There exist considerable differences of opinion in how to set restoration goals and how to define their success among conservation groups. Some urge active restoration (e.g. eradicating invasive animals to allow the native ones to survive) and others who believe that protected areas should have the bare minimum of human interference, such as rewilding. Ecosystem restoration has generated controversy. Skeptics doubt that the benefits justify the economic investment or who point to failed restoration projects and question the feasibility of restoration altogether. It can be difficult to set restoration goals, in part because, as Anthony Bradshaw claims, "ecosystems are not static, but in a state of dynamic equilibrium with restoration we aim for a moving target".

Some conservationists argue that, though an ecosystem may not be returned to its original state, the functions of the ecosystem (especially ones that provide services to us) may be more valuable than its current configuration. One reason to consider ecosystem restoration is to mitigate climate change through activities such as afforestation. Afforestation involves replanting forests, which remove carbon dioxide from the air. Carbon dioxide is a leading cause of global warming and capturing it would help alleviate climate change. Another example of a common driver of restoration projects in the United States is the legal framework of the Clean Water Act, which often requires mitigation for damage inflicted on aquatic systems by development or other activities.

Ecosystem restoration for the superb parrot on an abandoned railway line in Australia.

Forest restoration in action at the Buffelsdraai Landfill Site Community Reforestation Project in South Africa.

Restored prairie at the West Eugene Wetlands in Eugene, Oregon.

Ecological Restoration Challenges

Some view ecosystem restoration as impractical, partially because restorations often fall short of their goals. Hilderbrand point out that many times uncertainty (about ecosystem functions, species relationships, and such) is not addressed, and that the time-scales set out for 'complete' restoration are unreasonably short, while other critical markers for full-scale restoration are either ignored or abridged due to feasibility concerns. In other instances an ecosystem may be so degraded that abandonment (allowing a severely degraded ecosystem to recover on its own) may be the wisest option. Local communities sometimes object to restorations that include the introduction of large predators or plants that require disturbance regimes such as regular fires, citing threat to human habitation in the area. High economic costs can also be perceived as a negative impact of the restoration process.

Public opinion is very important in the feasibility of a restoration; if the public believes that the costs of restoration outweigh the benefits they will not support it.

Many failures have occurred in past restoration projects, many times because clear goals were not set out as the aim of the restoration, or an incomplete understanding of the underlying ecological framework lead to insufficient measures. This may be because, as Peter Alpert says, “people may not [always] know how to manage natural systems effectively”. Furthermore, many assumptions are made about myths of restoration such as carbon copy, where a restoration plan, which worked in one area, is applied to another with the same results expected, but not realized.

Science-practice Gap

One of the struggles for both fields is a divide between restoration ecology and ecological restoration in practice. Currently, many restoration practitioners as well as scientists feel that science is not being adequately incorporated into ecological restoration projects. In a 2009 survey of practitioners and scientists, the “science-practice gap” was listed as the second most commonly cited reason limiting the growth of both science and practice of restoration.

There are a variety of theories about the cause of this gap. However, it has been well established that one of the main issues is that the questions studied by restoration ecologists are frequently not found useful or easily applicable by land managers. Additionally many restoration ecology studies are carried out under controlled conditions and frequently at scales much smaller than actual restorations. Whether or not these patterns hold true in an applied context is often unknown. There is evidence that these small-scale experiments inflate type II error rates and differ from ecological patterns in actual restorations.

There is further complication in that restoration ecologists who want to collect large-scale data on restoration projects can face enormous hurdles in obtaining the data. Managers vary in how much data they collect, and how many records they keep. Some agencies keep only a handful of physical copies of data that make it difficult for the researcher to access. Many restoration projects are limited by time and money, so data collection and record keeping are not always feasible. However, this limits the ability of scientists to analyze restoration projects and give recommendations based on empirical data.

Contrasting Restoration Ecology and Conservation Biology

Restoration ecology may be viewed as a sub-discipline of conservation biology, the scientific study of how to protect and restore biodiversity. Ecological restoration is then a part of the resulting conservation movement.

Both restoration ecologists and conservation biologists agree that protecting and restoring habitat is important for protecting biodiversity. However, conservation biology is primarily rooted in population biology. Because of that, it is generally organized at the population genetic level and assesses specific species populations (i.e. endangered species). Restoration ecology is organized at the community level, which focuses on broader groups within ecosystems.

In addition, conservation biology often concentrates on vertebrate animals because of their salience and popularity, whereas restoration ecology concentrates on plants. Restoration ecology focuses on plants because restoration projects typically begin by establishing plant communities. Ecological restoration, despite being focused on plants, may also have "poster species" for individual ecosystems and restoration projects. For example, the Monarch butterfly is a poster species for conserving and restoring milkweed plant habitat, because Monarch butterflies require milkweed plants to reproduce. Finally, restoration ecology has a stronger focus on soils, soil structure, fungi, and microorganisms because soils provide the foundation of functional terrestrial ecosystems.

Forest Restoration

Forest restoration refers to the actions to re-instate ecological processes, which accelerate recovery of forest structure, ecological functioning and biodiversity levels towards those typical of climax forest, i.e. the end-stage of natural forest succession.

In places where forests have been lost or degraded, restoration or reforestation projects may be undertaken in order to guarantee or accelerate the recovery of forests. Objectives of forest restoration can range from economic incentives, to social or cultural values, to ecosystem services, to biodiversity conservation. The task of forest restoration can be a complex one, however, involving diverse ecological and social systems, which are not always fully accounted for or understood.

In the broad sense of the term, forest restoration is not new. When changes in patterns of human settlements and livelihood practices result in agricultural fields being abandoned, those areas can often naturally re-establish as forest, as natural succession processes occur. Active forms of forest restoration or reforestation, where specific actions are taken to ensure or accelerate re-establishment of forest cover in an area, have been used for utilitarian purposes throughout history. Examples of these include loblolly pine planting in the United States, or secondary spruce forests in central Europe, under threats of timber scarcity or interest in "improving" nature and providing employment. Out of these earlier utilitarian reforestation efforts emerged more intentional ecologically-focused restoration efforts, aiming to restore habitat, biodiversity, or ecological function of degraded ecosystems.

Forest restoration has become an area of increasing interest and activity in recent years, ranging from small-scale efforts by NGO's and communities to large-scale international commitments. One such example of a large-scale international commitment is the 20x20 Initiative, a country-led effort to restore 20 million hectares of land by 2020, in Latin America and the Caribbean. This effort is a part of the larger global Bonn Challenge, a global initiative launched in 2011 aiming to restore 150 million hectares worldwide by 2020. Also in recent years there has been an increased focus on Landscape Scale Initiatives, which aim to incorporate trees in a multi-use mosaic of various land uses, and take into account multiple stakeholders, land-uses and scales for decision making and adaptive management, allowing for human livelihoods as well as ecological integrity and ecosystem services across a landscape. The World Resources Institute's Global Restoration Initiative is home to "the Atlas of Forest and Landscape Restoration Opportunities," an interactive map that classifies land around the world and their potential for different kinds of restoration.

Natural Regeneration

Tree planting is not always essential to restore forest ecosystems. A lot can be achieved by studying how forests regenerate naturally, identifying the factors that limit regeneration and devising methods to overcome them. These can include weeding and adding fertilizer around natural tree seedlings, preventing fire, removing cattle and so on. This is "accelerated" or "assisted" natural regeneration. It is simple and cost-effective, but it can only operate on trees that are already present, mostly light-loving pioneer species. Such tree species are not usually those that comprise climax forests, but they can foster recolonization of the site by shade-tolerant climax forest tree species, via natural seed dispersal from remnant forest. Because this is a slow process, biodiversity recovery can usually be accelerated by planting some climax forest tree species, especially large-seeded, poorly dispersed species. It is not feasible to plant all the tree species that may have formerly grown in the original primary forest and it is usually unnecessary to do so, if the framework species method can be used.

Post-fire Regeneration

In large parts of the world, forest fires cover a heavy toll on forests. That can be because of provoked deforestation in order to substitute forests by crop areas, or in dry areas, because of wild fires occurring naturally or intentionally. A whole section of forest landscape restoration in linked to this particular problem, as in many cases, the net loss of ecosystem value is very high and can open the drop to an accelerated further degradation of the soil conditions through erosion and desertification. This indeed has dire consequences on both the quality of the habitats and their related fauna. Nevertheless, in some specific cases, wild fires do actually allow to increase the biodiversity index of the burnt area, in which case the Forest Restoration Strategies tend to look for a different land-use.

Forest Landscape Restoration

Forest landscape restoration (FLR) is defined as "a planned process to regain ecological integrity and enhance human well-being in deforested or degraded landscapes". It comprises tools and procedures to integrate site-level forest restoration actions with desirable landscape-level objectives, which are decided upon via various participatory mechanisms among stakeholders. The concept has grown out of collaboration among some of the world's major international conservation organizations including the International Union for Conservation of Nature (IUCN), the World Wide Fund for Nature (WWF), the World Resources Institute and the International Tropical Timber Organization (ITTO).

Aims

The concept of FLR was conceived to bring about compromises between meeting the needs of both humans and wildlife, by restoring a range of forest functions at the landscape level. It includes actions to strengthen the resilience and ecological integrity of landscapes and thereby keep future management options open. The participation of local communities is central to the concept, because they play a critical role in shaping the landscape and gain significant benefits from restored forest resources. Therefore, FLR activities are inclusive and participatory.

Desirable outcomes

The desirable outcomes of an FLR program usually comprise a combination of the following, depending on local needs and aspirations:

- identification of the root causes of forest degradation and prevention of further deforestation,
- positive engagement of people in the planning of forest restoration, resolution of land-use conflicts and agreement on benefit-sharing systems,
- compromises over land-use trade-offs that are acceptable to the majority of stakeholders,
- a repository of biological diversity of both local and global value,
- delivery of a range of utilitarian benefits to local communities including:
    - a reliable supply of clean water,
    - environmental protection particularly watershed services (e.g. reduced soil erosion, lower landslide risk, flood/drought mitigation etc.),
    - a sustainable supply of a diverse range of forest products including foods, medicines, firewood etc.,
    - monetary income from various sources e.g. ecotourism, carbon trading via the REDD+ mechanism and from payments for other environmental services (PES).

Activities

FLR combines several existing principles and techniques of development, conservation and natural resource management, such as landscape character assessment, participatory rural appraisal, adaptive management etc. within a clear and consistent evaluation and learning framework. An FLR program may comprise various forestry practices on different sites within the landscape, depending on local environmental and socioeconomic factors. These may include protection and management of secondary and degraded primary forests, standard forest restoration techniques such as "assisted" or "accelerated" natural regeneration (ANR) and the planting of framework tree species to restore degraded areas, as well as conventional tree plantations and agroforestry systems to meet more immediate monetary needs.

The IUCN hosts the Global Partnership on Forest Landscape Restoration, which co-ordinates development of the concept around the world.

In 2014, the Food and Agricultural Organization of the United Nations established the Forest and Landscape Restoration Mechanism. The Mechanism supports countries to implement FLR as a contribution to achieving the Bonn Challenge - the restoration of 150 million hectare of deforested and degraded lands by 2020 - and the Convention on Biological Diversity Aichi Biodiversity Targets - related to ecosystem conservation and restoration.

In partnership with the Global Mechanism of the United Nations Convention to Combat Desertification, FAO released two discussion papers on sustainable financing for FLR in 2015. *Sustainable Financing for Forest and Landscape Restoration: The Role of Public Policy Makers* provides recommendations and examples of FLR financing for countrues. *Sustainable Financing for Forest and Landscape Restoration - Opportunities, challenges and the way forward* provides an overview of funding sources and financial instruments available for FLR activities.

Example

The Ashland Forest Resiliency Stewardship Project (AFR) is a decade long, science-based project launched in 2010 with the intent of reducing severe wildfire risk, but also protecting water quality, old-growth forest, wildlife, people, property, and the overall quality of life within the Ashland watershed. The primary stakeholders in this cooperative restoration effort are the U.S. Forest Service, the City of Ashland, Lomaktsi Restoration Project, and the Nature Conservancy. The project was launched with initial funding from the Economic Recovery stimulus, and has more recently received funding from the Forest Service Hazardous Fuels program and the Joint Chiefs Landscape Restoration Partnerships program to back the project through 2016 .

Located in the dry forests of southern Oregon, the threat of wildfire is a reality for land managers and property owners alike. The boundaries of the city of Ashland intersect with the surrounding forest in what is referred to as the wildland-urban interface (WUI). Historically, the forests of this region experienced a relatively frequent fire return interval, which prevented buildup of heavy fuel loads. A century of fire exclusion and suppression on federal lands in the Pacific Northwest has led to increased forest density and fuel loads, and thus a more persistent threat of devastating wildfire.

The AFR project has implemented restoration techniques and prescriptions that aim to replicate the process of ecological succession in dry, mixed-conifer forests of the Pacific Northwest. The approach involves a combination of fuels reduction, thinning small-diameter trees, and carrying out prescribed burns. Priority is given to maintaining ecological function and complexity by retaining the largest and oldest trees, preserving wildlife habitat and riparian areas, and protecting erodible soils and maintaining slope stability.

Since its inception in 2010, the AFR project has completed restoration work on 4,000 of the 7,600 acres slated for the project. The project has provided educational experience to over 2,000 students and has benefitted the local community by creating jobs and providing workforce training. Currently, helicopter logging operations are thinning 1,100 acres of the watershed while controlled burning operations take place as air quality conditions allow.

Land Restoration

Land restoration includes all the techniques meant to restore the degraded land

Methods used in Land Restoration

Land restoration and revegetation is carried out through execution of corrective measures on land where the degradation has occurred. The current measures usually executed are summarized below:

On Cultivated Land

Agro-silvicultural Methods

These are practiced to restore the soil fertility. The traditional system throughout Africa, under rainfed conditions, is to restore the lost fertility through bush fallow system. Vegetation, mainly bushes, colonise the area naturally. After a resting period of 10-15 years in dry savanna land, the bushes are cleared and cropping is continued until the soil loses its fertility. The fallows are not protected and animals graze on them and they receive added fertilization from the dung of animals.

In the Sudan, Acacia Senegal is sown with the last agricultural crop or occurs from the coppice old trees which are not cut back in the last cropping. The successful seedlings are protected from browsing animals until the seedlings are out of the reach of animals. Then animals graze in these fallows making use of the grass and the pods of the gum trees and add to the soil fertility by their droppings. The gum trees continue for 15-20 years before they are cleared for cultivation of crops.

This is the system, as used to be practiced in the past. However, rotations for agricultural crops tended to be longer than the soil can support and the rotations of the bushes or gum trees tended to be shorter.

The technology used remained unchanged, save for the change from seeding to the use of seedlings, as the conditions in the drier parts of the savanna would not allow for seeding or coppicing of the older stock of trees. The farmer does the work and recently the Forest Service has been providing the seedlings free of charge, and the technical advice.

Shelterbelts

These are used to protect both irrigated and rainfed farms. Their main function, at present, is to protect valuable agricultural land and irrigation canals from creeping sands. This occurs in North Africa and Northern Sudan. Shelterbelts reduce wind velocity, improve the microclimate and increase livestock yields. Field investigations in dry areas show that crop production may be increased by as much as 300% while the increase in average years is often 30 to 50%. By ameliorating the microclimate, shelterbelts commonly improve fruit production by a third or more, and in windy years, the fruit value may be increased by over 75%. By protecting livestock from strong winds and heat, shelterbelts have a great influence on the production of flocks. Under protection, the lambing percentage is greater by about 30%. Excellent examples of shelterbelts have been established in Libya, Egypt, Syria and Iran during recent years. The farmers at large are, however, reluctant to sacrifice cultivated areas for them. The species used are mostly Eucalyptus, Casuarina sp., populus and prosopis. Seedlings are used for establishment.

Plantations on Seriously Degraded irrigated or Rainfed Crop Land

Salinization is of common occurrence in irrigated lands in some countries of the Middle East region. In Iraq, plantations of poplars and eucalyptus were established in some of these degraded lands to bring them back to production. The product is to be used to establish wood-based industries.

In the Gezira of the Sudan degraded lands, either because of high salinity or because of being in depressions and therefore liable to waterlogging, and those lands used as dumping ground for surplus water, are cut out of the cropping rotations. Eucalyptus microtheca plantations are established on such sites.

Also in the central clay plain of the Sudan where rainfed mechanized crop production is practiced, some of the areas that lost their fertility and could no longer be used for crop production were given to the Forest Department. Here, plantations of Acacia senegal were established where sowing and weeding were done by mechanical means. The objectives set for stocked ares are to restore fertility and to be kept as reserves of strategic value to stabilize the production of gum should the need arise.

On Rangeland

Degradation of rangeland is mainly due to overgrazing. Le Hourou (1981) summarized the actions of overgrazing as follows:

1. To reduce plant cover and biomass, mainly of perennials;
2. To reduce the number of annuals in the long term through lack of seed production;
3. To reduce the number of palatable species and individuals;
4. To replace palatable species by unpalatable ones;
5. To strongly reduce primary production and still more drastically secondary production;
6. To increase erosion as a result of a reduced plant cover.

Overgrazing is of common occurrence around watering points, such as wells, streams and rivers. Another cause of degradation is incidental or deliberate fires, particularly in the drier parts of the savanna in Africa (300-800 mm of annual rainfall). Fires cause direct loss of fodder through burning and reduce its quality.

Corrective measures carried out are:

- Total protection by fencing and thus restoring the vegetation after a few years;
- Seeding of palatable indigenous species, mostly of grass. Seedlings of various bushes and trees are planted to restore the range. These species are: Atriplex, Prosopis, Acacia and Salsola;
- Establishment of fire-lines in the dry savanna zone of Africa.

On forests and Woodlands

Here again, fires are considered a cause of degradation in dry savanna zones and fire-lines are established along the boundaries of Forest Reserves. Enrichment planting through establishment of plantations blocks are carried out. The objectives of such plantations are generally the supply of fuel wood and poles. In the wetter savanna formations, the objectives of these enrichment plantings change to timber production. Species used are Eucalyptus and Conifers, mainly: Cupressus lusitanica, Pinus patula and P. raddiata, P. halepensis, and other pines.

On Bare Land: Sand and Sand-dune Fixation

When erosion reaches an acute level, and where sand starts to move and threatens habitations, establishments, roads and agricultural land, sand-dune fixation is carried out. This occurs along the coasts of seas and oceans or in continental areas. The conventional method used is the establishment of hedges of stalks of dry grass and/or bushes to restrain temporarily sand movement until the dunes are planted with trees, shrubs and grasses. Supplementary irrigation is sometimes used. Eucalypts, pines, tamarisks and acacias are used in the higher latitudes along the Mediterranean in North Africa and the Middle East. In Sudan, Somalia and Mauritania prosopis and acacias are used.

Agroecological Restoration

Agroecological restoration is the practice of re-integrating natural systems into agriculture in order to maximize sustainability, ecosystem services, and biodiversity. This is one example of a way to apply the principles of agroecology to an agricultural system.

Farms cannot be restored to a purely natural state because of the negative economic impact on farmers, but returning processes, such as pest control to nature with the method of intercropping, allows a farm to be more ecologically sustainable and, at the same time, economically viable. Agroecological restoration works toward this balance of sustainability and economic viability because conventional farming is not sustainable over the long run without the integration of natural systems and because the use of land for agriculture has been a driving force in creating the present world biodiversity crisis. Its efforts are complementary to, rather than a substitute for, biological conservation.

biodiversity is just as important on farms and in fields as it is in deep river valleys or mountain cloud forests.

Agriculture creates a conflict over the use of land between wildlife and humans. Though the domestication of crop plants occurred 10,000 years ago, a 500% increase in the amount of pasture and crop land over the last three hundred years has led to the rapid loss of natural habitats. In recent years, the world community acknowledged the value of biodiversity in treaties, such as the 1992 landmark Convention on Biological Diversity.

Reintegration

The reintegration of agricultural systems into more natural systems will result in decreased yield and produce a more complex system, but there will be considerable gains in biodiversity and ecosystem services.

Biodiversity

The Food and Agriculture Organization of the United Nations estimates that more than 40% of earth's land surface is currently used for agriculture. And because so much land has been converted to agriculture, habitat loss is recognized as the driving force in biodiversity loss (FAO). This biodiversity loss often occurred in two steps, as in the American Midwest, with the introduction of mixed farming carried out on small farms and then with the widespread use of mechanized farming and monoculture beginning after World War II. The decline in farmland biodiversity can now be traced to changes in farming practices and increased agricultural intensity.

Increasing Heterogeneity

Heterogeneity (here, the diversity or complexity of the landscape) has been shown to be associated with species diversity. For example, the abundance of butterflies has been found to increase with heterogeneity. One important part of maintaining heterogeneity in the spaces between different fields is made up of habitat that is not cropped, such as grass margins and strips, scrub along field boundaries, woodland, ponds, and fallow land. These seemingly unimportant pieces of land are crucial for the biodiversity of a farm. The presence of field margins benefits many different taxa: the plants attract herbivorous insects, will which attract certain species of birds and those birds will attract their natural predators. Also, the cover provided by the no cropped habitat allows the species that need a large range to move across the landscape.

Monoculture

In the absence of cover, species face a landscape in which their habitat is greatly fragmented. The isolation of a species to a small habitat that it can't safely wander from can create a genetic bottleneck, decreasing the resilience of the particular population, and be another factor leading to the decline of the total population of the species. Monoculture, the practice of producing a single crop over a wide area, causes fragmentation. In conventional farming, monoculture, such as with rotations of corn and soybean crops planted in alternating growing seasons, is used so that very high yields can be produced. After the mechanization of farming, monoculture became a standard practice in corn-beans rotation, and had broad implications for the long-term sustainability and biodiversity of farms. Whereas organic fertilizers, had kept the soil's nutrients fixed to the ecosystem, the introduction of monoculture removed the nutrients and farmers compensated for that loss by using inorganic fertilizers. It is estimated that humans have doubled the rate of nitrogen input into the nitrogen cycle, mostly since 1975. As a result, the biological processes that controlled the way crops used the nutrients changed and the leached nitrogen from farmland soils has become a source of pollution.

Organic Farming

Organic farming is defined in different legal terms by different nations, but its main distinction from conventional farming is that it prohibits the use of synthetic chemicals in crop and livestock production. Often, it also includes diverse crop rotations and provides non-cropped habitat for insects that provide ecosystem services, such as pest control and pollination. However, it is merely encouraged that organic farmers follow those kinds of wildlife friendly practices, and as a result there is a great difference between the ecosystem services that similarly sized but distinctly managed organic farms provide.

1. Prohibition/reduced use of chemical pesticides and inorganic fertilizers is likely to have a positive impact through the removal of both direct and indirect negative effects on arable plants, invertebrates and vertebrates.
2. Sympathetic management of non-crop habitats and field margins can enhance diversity and abundance of arable plants, invertebrates, birds and mammals.
3. Preservation of mixed farming is likely to positively impact farmland biodiversity through the provision of greater habitat heterogeneity at a variety of temporal and spatial scales within the landscape.

Desert Greening

Desert greening is a concept where a number of methods are used to revitalize arid and semi-arid deserts. New methods are formulated every year. Here's a list of some of the revitalizing methods:

- Landscaping to reduce temperature, erosion, sandstorms, and evaporation;
- Greenhouse agriculture;
- Regeneration of salty, polluted, or degenerated soils;
- Permaculture - growing of plant communities;
- Planting trees and salt-loving plants such as Salicornia and Halophytes;
- Flood control;
- Seawater farming;
- Inland mariculture;
- Prevention of extensive grazing and firewood use;
- Training of local communities to care for plants, water, etc.

For example, the once lush Horqin Pasture in Mongolia has become a desert over time due to extensive grazing and deforestation brought about by a steady increase in the population of man and livestock. Lack of proper information on environment coupled with poor government support has led to the rapid deterioration of Horqin. As pastures disappear, farmers tend to move to other greenery, thus causing the Horqin Desert to expand at a rate of 10,000 km2 per year.

Water

Desert greening is more or less a function of water availability. If sufficient water for irrigation is at hand, any hot, cold , sandy or rocky desert can be greened. Water can be made available through saving, reuse, rainwater harvesting, desalination, or direct use of seawater for salt-loving plants. These different paths have unique features, i.e.: conserving water is a cheap solution. Reuse of treated water and the closing of cycles is the most efficient because closed cycles stand for unlimited and sustainable supply - rainwater management is a decentralized solution and applicable for inland areas - desalination is very secure as long as the primary energy for the operation of the desalination plant is available - Direct use of seawater for seawater agriculture is the most potent, only limited by the need for pumping up the water from sea-level.

Novel type of desalination is done with the Sahara Forest Project. This project uses solar stills for the generation of the freshwater. Another novel technique is cloud seeding, either by artificial means or through the action of cloud-seeding bacteria that live on vegetation (e.g. *Pseudomonas syringae*). Another, "atmospheric water generation" or air to water, uses dehumidification and is used by the military for potable water generation. However this technology uses 200 times more energy than desalination, making it unsuitable for large scale desert greening. An exception is the *Integrated Biotectural System*, which performs air-to-water in a greenhouse - a closed environment with optimized atmospheric conditions.

Lake Tuendae, an artificial pond at Zzyzx Desert Studies Center in the Mojave Desert.

Water Distribution

Once the (fresh or sea)water has been attained in centralized systems it must be distributed. This can be done using dug canals or in some instances aqueducts (which are both the least attractive option since they allow much water to be evaporated), troughs (as used in the Keita Project), earthenware piping (semi-open or closed) or even underground systems i.e. qanāt.

Depending on the method of distribution of the water, it can then be provided on different methods to the plants. A costly solution (used only on pipes) is drip irrigation. Other methods are the use of wadis (basically V-shaped ponds dug in the earth) or by simply planting the trees in holes inside/over the water pipe itself. The tree's roots can then suck the water straight from the water pipe (used in qanāt, hydroponics, ...) A similar technique can be done with semi-open pipes (i.e. dug throughs in the Keita Project).

Side Effects

The use of water is, however, not always without problems. Desert greening by the Helmand and Arghandab Valley Authority irrigation in Afghanistan significantly reduced the water flowing from the Helmand River into Lake Hamun and this, together with drought, was cited as a key reason for the severe damage to the ecology of Lake Hamun, much of which has degenerated since 1999 from a wetland of international importance into salt flats.

Trees

A main component of desert greening is the planting of trees. Trees store water, raise water from underlying aquifers, reduce evaporation after a rain, attract animals (and thereby fertility through feces), and they can cause more rain to fall (by temperature reduction and other effects), if the planted area is large enough.

Example

The soil of the Thar Desert in India remains dry for much of the year and is prone to wind erosion. High velocity winds blow soil from the desert, depositing some on neighboring fertile lands, and causing shifting sand dunes within the desert, which buries fences and blocks roads and railway tracks. A permanent solution to this problem of shifting sand dunes can be provided by planting appropriate species on the dunes to prevent further shifting and planting windbreaks and shelterbelts. These solutions also provide protection from hot or cold and desiccating winds and the invasion of sand. The Rajasthan Canal system in India is the major irrigation of the Thar Desert and is intended to reclaim it and to check spreading of the desert to fertile areas.

Prevention of shifting sand dunes is accomplished through plantations of *Acacia tortilis* near Laxmangarh town. There are few local tree species suitable for planting in the desert region and these are slow growing. The introduction of exotic tree species in the desert for plantation has become necessary. Many species of Eucalyptus, Acacia, Cassia and other genera from Israel, Australia, US, Russia, Zimbabwe, Chile, Peru, and Sudan have been tried in the Thar Desert. *Acacia tortilis* has proved to be the most promising species for desert greening. The jojoba is another promising species of economic value which has been found suitable for planting in these areas.

Sundrop Farms launched a greenhouse in 2016 to produce 15,000 tonnes of tomatoes using only desert soil and desalinated water piped from Spencer Gulf.

Island Restoration

The ecological restoration of islands, or island restoration, is the application of the principles of ecological restoration to islands and island groups. Islands, due to their isolation, are home to many of the world's endemic species, as well as important breeding grounds for seabirds and some marine mammals. Their ecosystems are also very vulnerable to human disturbance and particularly to introduced species, due to their small size. Island groups such as New Zealand and Hawaii have undergone substantial extinctions and losses of habitat. Since the 1950s several organisations and government agencies around the world have worked to restore islands to their original states; New Zealand has used them to hold natural populations of species that would otherwise be unable to survive in the wild. The principal components of island restoration are the removal of introduced species and the reintroduction of native species.

Islands, Endemism and Extinction

Isolated islands have been known to have greater levels of endemism since the 1970s when the theory of Island biogeography, formulated by Robert MacArthur and E.O. Wilson was developed. This higher occurrence of endemism is because isolation limits immigration of new species to the island, allowing new species to evolve separately from others on the mainland. For example, 71% of New Zealand's bird species (prior to human arrival) were endemic. As well as displaying greater levels of endemism, island species have characteristics that make them particularly vulnerable to human disturbance.

Many island species evolved on small islands, or even restricted habitats on small islands. Small populations are vulnerable to even modest hunting, and restricted habitats are vulnerable to loss or modification of said habitat. More importantly, island species are often ecologically naive, that is they have not evolved alongside a predator, or have lost appropriate behavioural responses to predators. This often resulted in flightlessness, or unusual levels of tameness. This made many species susceptible to hunting (it is thought, for example, that moas were hunted to extinction in a few short generations) and to predation by introduced species. Some, such as the dodo, are thought to have become extinct because of the pressure of both humans and introduced animals. One estimate of birds in the Pacific islands puts the extinctions at 2000 species. Between 40 and 50% of the bird species of New Zealand have become extinct since 200 AD.

Takahe have bred after translocation to restored islands, like these on Kapiti Island.

The field of island restoration is usually credited with having been started in New Zealand in the 1960s, but other smaller projects, such as the restoration of Nonsuch Island in Bermuda (which began in 1962) have been going on for almost as long. Nevertheless, the program undertaken by the Department of Conservation (DOC) is one of the largest in the world. It began on Cuvier Island, where ecologists removed stock, goats, feral cats and finally, in 1993, Pacific rats. The success of the project resulted in similar projects around New Zealand. The advantages to the DOC were considerable; in addition to protecting species endemic to smaller islands, like the magenta petrel, islands near the mainland, once restored, could act as habitat for species of birds that were unable to survive on the mainland. Species like the takahe, where the remaining wild population was at considerable risk from feral cats and dogs, could be moved to these islands to safeguard the species.

Eradication of Introduced Alien Species

One important aspect of island restoration is the removal of invasive alien species. Since these species are most often the reason that native fauna and flora is threatened, their removal is essential to the restoration project. From 1673 until 2009, 786 successful invasive vertebrate eradication have been recorded and in the last few decades the frequency of eradications and the size of islands from which invasive vertebrates have been eradicated has increased.

Islands are particularly suitable for restoration as once cleared of an introduced species they can be kept cleared of these species by virtue of being an island. Species removal is intensive and expensive, and methods used must be carefully chosen as to not create too much impact on non-target species. Feral cats, goats and three species of rats are among the most damaging species introduced to islands (Moors & Atkinson 1984). The differences in size, lifestyle and behaviour preclude the use of the same techniques for all of them, but with many species a range of techniques needs to be used in order to ensure success. Larger animals, such as goats and pigs, can be effectively hunted; in the case of Round Island, in Mauritius, all the goats were eliminated by a single marksman. On larger islands ecologists use a Judas goat, where a radio collared goat is released into the wild. This goat is then followed and groups it joins are removed. To remove cats a combination of techniques is needed: hunting, trapping and poisoning. Cats are more difficult to hunt than goats and pigs, requiring the use of experienced hunters and night hunting.

Trapping is ineffective for rats, given their sheer numbers, and the only method that works is poisoning, which can be delivered into the field by broadcasting (by hand or from the air) or by the maintenance of bait stations. This method has been employed around the world, in the Falkland Islands, in the tropical Pacific, and off New Zealand, where over 40 islands have been cleared. This method is not without problems, especially if the rats share the island with other, native species of rodent that might take the bait as well, as has happened on Anacapa Island in the Channel Islands and Rat Island (Hawadax) in the Aleutian archipelago. In the Pacific poison intended for rats was taken by land crabs instead; the crabs were not affected by the poison but frustrated attempts to clear the rats.

The removal of invasive weeds is, in most cases, more difficult than removing animal species. One such eradication was that of sandbur, *Cencrus echinatus*, an introduced grass on Laysan. The grass, introduced to Laysan around 1961, had taken over 30% of the island by 1991, displaced the native bunchgrass, and reduced the breeding habitat of two endemic threatened species, the Laysan duck and Laysan finch, as well as those of several seabirds. The removal took ten years, with controlled spraying for the first year, then individual removal of plants, then, when few plants were being found, sifting of the sands around plants to remove seeds. The cost of the eradication program was $150,000/year.

Strategies

There are three strategies to reduce the numbers of the introduced animals in order to lighten the problems caused by them: exclusion, control, and eradication. Exclusion is removal of introduced species in limited areas and focused on a local impact. Control has a meaning of mitigation by reducing the numbers of introduced species down to ecologically and economically less harmful level. Since it is not complete wiping out strategy, this has to be taken constantly and repeatedly. This strategy is said to be feasible but less cost effective due to its incompleteness. Eradication is the complete removal of all the individuals of the population, down to the last potentially reproducing individual, or the reduction of their population density below sustainable levels . For islands, it is usually the best strategy as it gives permanent effects, which potentially means most cost effective. However, it is still logistically and economically high cost. Also, this strategy is sometimes hard to accomplish depending on the environment of island and the alien species.

Methods

Fencing is used for excluding the alien mammals. This method intends to limit the area for the mammals before control or eradication. Shooting is often used for large animals to control them. This method has meanings of gaining food supply and recreation. It requires accessibility in the field and experience for the hunters. Also, to maintain the effect, it requires good amount of manpower for long time, so it can be costly. Trapping is designed for medium-sized mammals which are hard to shoot at. This method is usually only for reduction not eradication. The advantage of this method is that it can be selective since it is possible to exclude or reduce the possibility of trapping native animals. However, trapping requires limited area to cover and limited population to capture.

Poisoning is very effective for small species. However, there are some disadvantages of this method. It is possible that untargeted animals take the poison. It is also necessary to consider the secondary poisoning that other animals are affected by poisoned species. This method can be costly if the area to cover is large.

Pathogen Introduction (parasitism) is one of the biological methods to eradicate alien species. It is very effective for limited species, but the viruses and bacteria need to be specified clearly for use.

Predator introduction is another biological method to eradicate the introduced species. It is less costly and environmentally clean, but it can cause greater problems, because it is possible that the introduced predator targets the native animals instead of alien species and its existence and its parasites can be new problems.

Competitor introduction is also a biological method to eradicate introduced carnivores. It can be very effective with good amount of information. In the end, the competitor has to be removed as well. Virus vectored immune-contraception is one of the newest method that is to infect introduced animals with genetically engineered viruses. This method is considered environmentally clean, low cost, selective, and ethical. However, it is not fully operational and the effect comes slowly.

Restoration of Former Habitat

In many cases the removal of introduced species is sufficient to allow a return to a pre-disturbance state, but generally active management, often in the form of replanting native flora and reintroduction of extirpated fauna is needed to achieve restoration goals. Planting of native species helps to replenish species that were either grazed or out competed. Species of animal can be translocated either from existing populations, or from captive bred populations. These reintroductions need to be carefully managed, particularly in the case of endangered species, where the potential benefits need to be weighed against the possibility of failure. Not all translocations succeed, and it may be necessary to help the reintroduced animals along with supplementary feeding or other kinds of management.

One other important aspect of restoration is prevention that is, keeping invasive species from returning to a cleared island. This can be achieved by restricting access to the island in question (reducing possible instances of invasion) to more stringent quarantine methods. For example, in order to prevent invasive weeds from returning to Laysan, people working on the island must bring entirely new clothes to the island, which must be frozen prior to arrival.

Opposition to Island Restoration

Prior to the initial efforts to remove rats from New Zealand's offshore islands there was a great deal of skepticism as to the feasibility of island restoration amongst ecologists and conservation workers. However, as the techniques have improved and larger islands have been restored, most of the initial criticisms from within the field have been dropped, in particular as the costs of eradication are often much lower than continuous pest control. Outside of the field of conservation there has been some opposition from other interested groups, particularly from the animal rights movement, which contends that the welfare of the pests in question is not adequately addressed in island restoration plans. Because a broad spectrum of pest removal techniques needs to be used, including leg traps, animal rights campaigners accuse ecologists of cruelty, and indifference to non-targeted species that also take bait or are trapped, and suggest that more humane methods such as capture and sterilization be used instead (something those working in island restoration contend would be too expensive, and potentially ineffective as in Kangaroo Island koalas.). Some also defend the rights of the introduced species to exist as well. Others, including scientists affiliated with the animal rights movement, accept that when the choice is between the future of a species and a population of pests, the future of a species must take priority (with the caveat that the extermination is conducted as humanely as possible). Opposition to island restoration has not led to the abandonment of many projects but has delayed several, particularly through court action. Groups sometimes adopt different approaches; opponents of hedgehog removal in the Outer Hebrides offered bounties for live hedgehogs removed from the islands and relocated in their natural habitat. Invasive plants can also generate strong feelings. The removal of *Eucalyptus* trees from Angel Island in San Francisco Bay faced considerable opposition.

Island Restoration Projects

Island restoration has been attempted in many countries since the 1960s, and has met with varying degrees of success. The following examples highlight some of the factors that influence projects. A comprehensive list of projects can be found on the Database of Island Invasive Species Eradications.

Round Island

Round Island is a tropical island 22.5 kilometres north of Mauritius, with an area of 1.69 square kilometres and a maximum elevation of 280 metres. Compared to other Mascarene islands and islets, Round Island remained relatively pristine until goats and rabbits were introduced in the 19th century (goats were introduced between 1846 and 1868, whilst rabbits were present in large numbers before 1810). These herbivores, along with other factors, such as sporadic logging and frequent cyclones, led to the eventual dwindling of the island's forests and dependent fauna. In turn, this led to soil erosion by wind and rain, impeding forest regeneration, acting as a positive feedback mechanism that caused rapid deforestation of the island. In 1957, however, Round Island was officially made a nature reserve, and in 1979 and 1986 was rid of goats (shooting) and rabbits (poisoning), respectively, after several unsuccessful attempts.

Since removal of the introduced herbivores, the Round Island plant community has recovered dramatically. This is especially stark for three of the endemic tree species, *Latania loddigesii*, *Pandanus vandermeerschii* and *Hyophorbe lagenicaulis*, which constituted a large portion of the Round Island forest historically. This has led to six reptile species (five of which were critically endangered) recovering in tandem with the plant community; these are the skinks *Leiolopisma telfaririi* and *Scelotes bojerii*, the geckos *Phelsuma guentheri*, *P. ornata* and *Nactus serpensinsula*, and the snake *Casarea dussumerii*.

The Round Island restoration program represents one of the longest-running projects of its kind in the world, since its conservation status was confirmed in 1957. Much of the current conservation work on Round Island is conducted by the Mauritian Wildlife Foundation and Durrell Wildlife Conservation Trust and revolves around maintaining soil levels, reforestation of the island, and eradication of remaining invasive plants and invertebrates.

Aleutian Islands

Arctic fox with least auklet.

The Aleutian Islands, prior to the 18th century, lacked any terrestrial predators, but from the mid-18th century Arctic foxes were introduced to act as a source for the fur trade, a practice that continued into the early 20th century. This introduction decimated the birds of the chain, particularly seabirds like the whiskered auklet. The reduction in seabirds, in turn, had effects on the ecology of the islands, as many of the plants were dependent on the guano from nesting birds acting as a fertilizer. In the early 1950s managers of the Aleutian Islands Reservation became aware of the damage, and an eradication program began. Since then over 80 islands have been cleared of non-native foxes (only six islands remain) and bird populations have rebounded. Whiskered auklets, which numbered 25,000 in 1974, had increased to 116,000 in 2003.

Campbell Island

Campbell Island is a sub Antarctic island 700 km south of New Zealand that became infested with rats in the 19th century. Several endemic birds, including the Campbell teal and Campbell snipe, only survived on small rocky islets just off the island, and the populations were perilously low. Several teals were taken into captivity for ex-situ conservation, but once they had bred in captivity there was no-where else to return them to until the island was cleared of rats. The DOC's plan to remove rats from the island was one of the most ambitious attempted, as the island was so remote, the rat populations had the highest density of rats anywhere in the world, the weather treacherous and, at 113 km^2, it was the largest island at that point where eradication had been attempted. The poison had to be dropped in the winter, to minimize disturbance to nesting seabirds and reduce the chance of bird strike for the pilots. After several experiments, the eradication began in 2001. In 2003 trackers with dogs were unable to find any rats. Soon after the island was cleared it was possible to return the teals to the island. Snipe have self-reintroduced to the island and have begun breeding.

South Georgia

Rats, brought to South Georgia Island as stowaways on sealing and whaling ships in the late 18th century, have caused much damage to native wildlife, destroying tens of millions of ground-nesting birds' eggs and chicks. While previously the island's glaciers formed a natural barrier to the spread of rats, these glaciers are now slowly melting as the climate warms. In 2011, scientists instituted a four-year programme to entirely eradicate the rats and mice, in what would be by far the largest rodent eradication attempt in the world to date. The project was led by zoologist Anthony Martin of The University of Dundee who stated, "This is a man-induced problem and it's about time that man put right earlier errors." In July 2013, the success of the main phase of the extermination of the rats, which took place in May that year, was announced. 180 tonnes of rat poison, brodifacoum, were dropped over 70% of the island, in what was the world's largest ever operation of this kind. Another 95t of rat poison was planned to be dropped by three helicopters in January 2015. In June 2015 the eradication programme concluded, apparently successfully, with the island believed "very likely" to be rat free. Monitoring will continue for a further two or three years.

Habitat Management

Habitat management is meant to ensure the preservation of game animal habitats. Habitat management measures deliver long-term effects. In forest management, emphasis can be given to measures that benefit game animals.

Volunteer work is valuable in the management of game animal habitats.

In the management of cervid habitats, the animals' access to food during the winter should be ensured. When the density of broad-leaved trees is appropriate, cervids do not damage coniferous trees grown in the forest. The animals' large home ranges are a challenge in the management of cervid habitats.

Key elements in the management of forest grouse habitats include food and cover. However, different grouse species have different habitat needs. Capercaillie prefers mature forests further away from human activities. Black grouse thrives better in forests that are in earlier stages of succession. Hazel grouse needs alder forest to feed on and spruce forest to provide cover. Forests flanking peatlands are important habitats for willow grouse.

In the management of field game habitats, the aim is controlled "unmanagement". Moderate autumn ploughing may enable field game to have more cover and improve the status of the population in the area.

The most important waterfowl habitats include their breeding areas and staging areas during migration. Wintering areas play a more significant role in other parts of Europe. For waterfowl, important game management measures include establishing and restoring wetlands.

Values of Natural Habitat

The natural environment is a source for a wide range of resources that can be exploited for economic profit, for example timber is harvested from forests and clean water is obtained from natural streams. However, land development from anthropogenic economic growth often causes a decline in the ecological integrity of nearby natural habitat. For instance, this was an issue in the northern rocky mountains of the USA.

However, there is also economic value in conserving natural habitats. Financial profit can be made from tourist revenue, for example in the tropics where species diversity is high, or in recreational sports which take place in natural environments such as hiking and mountain biking. The cost of repairing damaged ecosystems is considered to be much higher than the cost of conserving natural ecosystems.

Measuring the worth of conserving different habitat areas is often criticized as being too utilitarian from a philosophical point of view.

Biodiversity

Habitat conservation is important in maintaining biodiversity, an essential part of global food security. There is evidence to support a trend of accelerating erosion of the genetic resources of agricultural plants and animals. An increase in genetic similarity of agricultural plants and animals means an increased risk of food loss from major epidemics. Wild species of agricultural plants have been found to be more resistant to disease, for example the wild corn species Teosinte is resistant to 4 corn diseases that affect human grown crops. A combination of seed banking and habitat conservation has been proposed to maintain plant diversity for food security purposes.

Classifying Environmental Values

Pearce and Moran outlined the following method for classifying environmental uses:

- Direct extractive uses: e.g. timber from forests, food from plants and animals.
- Indirect uses: e.g. ecosystem services like flood control, pest control, erosion protection.
- Optional uses: future possibilities e.g. unknown but potential use of plants in chemistry/ medicine.
- Non-use values:
    - Bequest value (benefit of an individual who knows that others may benefit from it in future).
    - Passive use value (sympathy for natural environment, enjoyment of the mere existence of a particular species).

Impacts

Natural Causes

Habitat loss and destruction can occur both naturally and through anthropogenic causes. Events leading to natural habitat loss include climate change, catastrophic events such as volcanic explosions and through the interactions of invasive and non-invasive species. Natural climate change, events have previously been the cause of many widespread and large scale losses in habitat. For example, some of the mass extinction events generally referred to as the "Big Five" have coincided with large scale such as the Earth entering an ice age, or alternate warming events. Other events in the big five also have their roots in natural causes, such as volcanic explosions and meteor collisions. The Chicxulub impact is one such example, which has previously caused widespread losses in habitat as the Earth either received less sunlight or grew colder, causing certain fauna and flora to flourish whilst others perished. Previously known warm areas in the tropics, the most sensitive habitats on Earth, grew colder, and areas such as Australia developed radically different flora and fauna to those seen today. The big five mass extinction events have also been linked to sea level changes, indicating that large scale marine species loss was strongly influenced by loss in marine habitats, particularly shelf habitats. Methane-driven oceanic eruptions have also been shown to have caused smaller mass extinction events.

Human Impacts

Humans have been the cause of many species' extinction. Due to humans' changing and modifying their environment, the habitat of other species often become altered or destroyed as a result of human actions. Even before the modern industrial era, humans were having widespread and major effects on the environment. A good example of this is found in Aboriginal Australians and Australian megafauna. Aboriginal hunting practices, which included burning large sections of forest at a time, eventually altered and changed Australia's vegetation so much that many herbivorous megafauna species were left with no habitat and were driven into extinction. Once herbivorous megafauna species became extinct, carnivorous megafauna species soon followed. In the recent past, humans have been responsible for causing more extinction within a given period of time than ever before. Deforestation, pollution, anthropogenic climate change and human settlements have all been driving forces in altering or destroying habitats. The destruction of ecosystems such as rainforests has resulted in countless habitats being destroyed. These biodiversity hotspots are home to millions of habitat specialists, which do not exist beyond a tiny area. Once their habitat is destroyed, they cease to exist.This destruction has a follow-on effect, as species which coexist or depend upon the existence of other species also become extinct, eventually resulting in the collapse of an entire ecosystem. These time-delayed extinctions are referred to as the extinction debt, which is the result of destroying and fragmenting habitats. As a result of anthropogenic modification of the environment, the extinction rate has climbed to the point where the Earth is now within a sixth mass extinction event, as commonly agreed by biologists. This has been particularly evident, for example, in the rapid decline in the number of amphibian species worldwide.

Approaches and Methods of Habitat Conservation

Determining the size, type and location of habitat to conserve is a complex area of conservation biology. Although difficult to measure and predict, the conservation value of a habitat is often a reflection of the quality (e.g. species abundance and diversity), endangerment of encompassing ecosystems, and spatial distribution of that habitat.

Identifying Priority Habitats for Conservation

Habitat conservation is vital for protecting species and ecological processes. It is important to conserve and protect the space/ area in which that species occupies. Therefore, areas classified as 'biodiversity hotspots', or those in which a flagship, umbrella, or endangered species inhabits are often the habitats that are given precedence over others. Species that possess an elevated risk of extinction are given the highest priority and as a result of conserving their habitat, other species in that community are protected thus serving as an element of gap analysis. In the United States of America, a Habitat Conservation Plan (HCP) is often developed to conserve the environment in which a specific species inhabits. Under the U.S. Endangered Species Act (ESA) the habitat that requires protection in an HCP is referred to as the 'critical habitat'. Multiple-species HCPs are becoming more favourable than single-species HCPs as they can potentially protect an array of species before they warrant listing under the ESA, as well as being able to conserve broad ecosystem components and processes. As of January 2007, 484 HCPs were permitted across the United States, 40 of which covered 10 or more species. The San Diego Multiple Species Conservation Plan (MSCP) encompasses 85 species in a total area of 26,000-km2. Its aim is to protect the habitats of multiple species and overall biodiversity by minimizing development in sensitive areas.

HCPs require clearly defined goals and objectives, efficient monitoring programs, as well as successful communication and collaboration with stakeholders and land owners in the area. Reserve design is also important and requires a high level of planning and management in order to achieve the goals of the HCP. Successful reserve design often takes the form of a hierarchical system with the most valued habitats requiring high protection being surrounded by buffer habitats that have a lower protection status. Like HCPs, hierarchical reserve design is a method most often used to protect a single species, and as a result habitat corridors are maintained, edge effects are reduced and a broader suite of species are protected.

Amount of Habitat Needed

A range of methods and models currently exist that can be used to determine how much habitat is to be conserved in order to sustain a viable population, including Resource Selection Function and Step Selection models. Modelling tools often rely on the spatial scale of the area as an indicator of conservation value. There has been an increase in emphasis on conserving few large areas of habitat as opposed to many small areas. This idea is often referred to as the "single large or several small", SLOSS debate, and is a highly controversial area among conservation biologists and ecologists. The reasons behind the argument that "larger is better" include the reduction in the negative impacts of patch edge effects, the general idea that species richness increases with habitat area and the ability of larger habitats to support greater populations with lower extinction probabilities. Noss & Cooperrider support the "larger is better" claim and developed a model that implies areas of habitat less than 1000ha are "tiny" and of low conservation value. However, Shwartz suggests that although "larger is better", this does not imply that "small is bad". Shwartz argues that human induced habitat loss leaves no alternative to conserving small areas. Furthermore, he suggests many endangered species which are of high conservation value, may only be restricted to small isolated patches of habitat, and thus would be overlooked if larger areas were given a higher priority. The shift to conserving larger areas is somewhat justified in society by placing more value on larger vertebrate species, which naturally have larger habitat requirements.

Examples of Current Conservation Organizations

The Nature Conservancy

Since its formation in 1951 The Nature Conservancy has slowly developed into one of the world's largest conservation organizations. Currently operating in over 30 countries, across 5 continents worldwide, The Nature Conservancy aims to protect nature and its assets for future generations. The organization purchases land or accepts land donations with the intension of conserving its natural resources. In 1955 The Nature Conservancy purchased its first 60-acre plot near the New York/Connecticut border in the United States of America. Today the Conservancy has expanded to protect over 119 million acres of land, 5,000 river miles as well as participating in over 1000 marine protection programs across the globe . Since its beginnings The Nature Conservancy has understood the benefit in taking a scientific approach towards habitat conservation. For the last decade the organization has been using a collaborative, scientific method known as 'Conservation by Design'. By collecting and analyzing scientific data The Conservancy is able to holistically approach the protection of various ecosystems. This process determines the habitats that need protection, specific elements that should be conserved as well as monitoring progress so more efficient practices can be developed for the future.

The Nature Conservancy currently has a large number of diverse projects in operation. They work with countries around the world to protect forests, river systems, oceans, deserts and grasslands. In all cases the aim is to provide a sustainable environment for both the plant and animal life forms that depend on them as well as all future generations to come.

World Wildlife Fund (WWF)

The World Wildlife Fund (WWF) was first formed in after a group of passionate conservationists signed what is now referred to as the Morges Manifesto. WWF is currently operating in over 100 countries across 5 continents with a current listing of over 5 million supporters. One of the first projects of WWF was assisting in the creation of the Charles Darwin Research Foundation which aided in the protection of diverse range of unique species existing on the Galápagos' Islands, Ecuador. It was also a WWF grant that helped with the formation of the College of African Wildlife Management in Tanzania which today focuses on teaching a wide range of protected area management skills in areas such as ecology, range management and law enforcement. The WWF has since gone on to aid in the protection of land in Spain, creating the Coto Doñana National Park in order to conserve migratory birds and The Democratic Republic of Congo, home to the world's largest protected wetlands. The WWF also initiated a debt-for-nature concept which allows the country to put funds normally allocated to paying off national debt, into conservation programs that protect its natural landscapes. Countries currently participating include Madagascar, the first country to participate which since 1989 has generated over $US50 million towards preservation, Bolivia, Costa Rica, Ecuador, Gabon, the Philippines and Zambia.

Rare Conservation

Rare has been in operation since 1973 with current global partners in over 50 countries and offices in the United States of America, Mexico, the Philippines, China and Indonesia. Rare focuses on the human activity that threatens biodiversity and habitats such as overfishing and unsustainable agriculture. By engaging local communities and changing behaviour Rare has been able to launch campaigns to protect areas in most need of conservation. The key aspect of Rare's methodology is their "Pride Campaign's". For example, in the Andes in South America, Rare has incentives to develop watershed protection practices. In the Southeast Asia's "coral triangle" Rare is training fishers in local communities to better manage the areas around the coral reefs in order to lessen human impact. Such programs last for three years with the aim of changing community attitudes so as to conserve fragile habitats and provide ecological protection for years to come.

WWF Netherlands

WWF Netherlands, along with ARK Nature, Wild Wonders of Europe and Conservation Capital have started the Rewilding Europe project. This project intents to rewild several areas in Europe.

Floodplain Restoration

A floodplain is a flat or nearly flat lowland bordering a stream or river that experiences occasional or periodic flooding. It includes the floodway, which consists of the stream channel and adjacent areas that carry flood flows, and the flood fringe, which are areas covered by the flood, but which do not experience a strong current. The floodplain corridor acts as the "right-ofway" for a stream and functions as an integral part of the stream ecosystem. Floodplains perform important natural functions, including temporary storage of floodwaters, moderation of peak flows, maintenance of water quality, groundwater recharge, and prevention of erosion. Floodplains also provide habitat for wildlife, recreational opportunities, and aesthetic benefits.

Ideal floodplain conditions are rarely encountered in urban or disturbed watersheds; many floodplain functions have been lost to agricultural uses and more recently to urban and suburban development. Floodplain restoration is the process of fully or partially restoring a stream's access to its floodplain to return those valuable floodplain functions.

There are multiple types of floodplain restoration:

- Hydrologic: Reconnecting the stream to the floodplain and restoring the stream's natural hydrology.
- Vegetative: Removing invasive species and replanting native plant communities appropriate to the site and conditions.
- Habitat Restoration: Installing structures to improve wildlife habitat. Habitat is also gained through re-planting native plant communities.

Several examples of restored and reconnected floodplains are displayed in Figures through.

Figure: These are two examples of streams that have been disconnected from their floodplains.

On the left, upstream development has increased stream flows, resulting in channel erosion and down cutting. The entrenched stream no longer overflows onto its floodplain. On the right, the floodplain was filled in and raised during development, cutting off the stream's access to its floodplain.

Figure: Examples of riparian buffer enhancement. In both cases, additional riparian trees and shrubs have been planted to expand the width of the forested buffer.

Figure: Floodplain re-vegetation, before (left) and after (right).

Figure: A stream restoration project with floodplain benching. On both sides of the restored stream, a bench has been graded at the bankfull elevation, providing room to dissipate high flow events and opportunities for riparian habitat creation.

References

- Simberloff, Daniel (January 1996). "Lawton, J. H. and May, R. M. (Eds.). Extinction Rates. 1995. Oxford University Press, Oxford. xii + 233 pp. ISBN 0-19-854829. X. Price: f17.95."Journal of Evolutionary Biology. 9 (1): 124–126. doi:10.1046/j.1420-9101.1996.t01-1-9010124.x. ISSN 1010-061X
- What-is-ecological-restoration: erabg.org, Retrieved 15 May 2018
- Hughes, A. Randall; Inouye, Brian D.; Johnson, Marc T. J.; Underwood, Nora; Vellend, Mark (2008-04-08). "Ecological consequences of genetic diversity". Ecology Letters. 11(6): 609–623. doi:10.1111/j.1461-0248.2008.01179.x. ISSN 1461-023X
- "Targeted habitat restoration can reduced extinction rates in fragrmented forests". National Academy of Sciences. Retrieved 19 December 2017
- Restoration-ecology, historians-european-biographies, literature-and-arts: encyclopedia.com, Retrieved 12 March 2018
- Paterson, Owen (27 November 2012). "Owen Paterson's speech to Royal Society on Natural Capital Committee". gov.uk. Retrieved June 4, 2018
- Jordan, III, W. R. (2011). Making nature whole : a history of ecological restoration. Washington, DC: Island Press. ISBN 9781610910422. OCLC 750183084

- Forest-restoration-reforestation, conservation: globalforestatlas.yale.edu, Retrieved 25 April 2018
- Kettenring, Karin M.; Adams, Carrie Reinhardt (2011-08-01). "Lessons learned from invasive plant control experiments: a systematic review and meta-analysis". Journal of Applied Ecology. 48 (4): 970–979. doi:10.1111/j.1365-2664.2011.01979.x. ISSN 1365-2664

Permissions

All chapters in this book are published with permission under the Creative Commons Attribution Share Alike License or equivalent. Every chapter published in this book has been scrutinized by our experts. Their significance has been extensively debated. The topics covered herein carry significant information for a comprehensive understanding. They may even be implemented as practical applications or may be referred to as a beginning point for further studies.

We would like to thank the editorial team for lending their expertise to make the book truly unique. They have played a crucial role in the development of this book. Without their invaluable contributions this book wouldn't have been possible. They have made vital efforts to compile up to date information on the varied aspects of this subject to make this book a valuable addition to the collection of many professionals and students.

This book was conceptualized with the vision of imparting up-to-date and integrated information in this field. To ensure the same, a matchless editorial board was set up. Every individual on the board went through rigorous rounds of assessment to prove their worth. After which they invested a large part of their time researching and compiling the most relevant data for our readers.

The editorial board has been involved in producing this book since its inception. They have spent rigorous hours researching and exploring the diverse topics which have resulted in the successful publishing of this book. They have passed on their knowledge of decades through this book. To expedite this challenging task, the publisher supported the team at every step. A small team of assistant editors was also appointed to further simplify the editing procedure and attain best results for the readers.

Apart from the editorial board, the designing team has also invested a significant amount of their time in understanding the subject and creating the most relevant covers. They scrutinized every image to scout for the most suitable representation of the subject and create an appropriate cover for the book.

The publishing team has been an ardent support to the editorial, designing and production team. Their endless efforts to recruit the best for this project, has resulted in the accomplishment of this book. They are a veteran in the field of academics and their pool of knowledge is as vast as their experience in printing. Their expertise and guidance has proved useful at every step. Their uncompromising quality standards have made this book an exceptional effort. Their encouragement from time to time has been an inspiration for everyone.

The publisher and the editorial board hope that this book will prove to be a valuable piece of knowledge for students, practitioners and scholars across the globe.

Index